STUDIO ACOUSTICS

by

Michael Rettinger

Consultant on Acoustics

Chemical Publishing Co., Inc.
New York, N.Y.

Chemical Publishing Co., Inc.

ISBN 0-8206-0283-3
ISBN 0-8206-0376-7

Printed in the United States of America

FOREWORD

Sound recording studios are often built like showcases, either to attract clientele or to provide a distinctive image for the industry. They are, thus, like people, in that no two of them are alike. Yet, all such structures have to have certain common acoustic elements if they are to function to the best artistic and economic advantages. The enclosures must be sufficiently quiet, exhibit proper reverberatory conditions (often required to be adjustable), be devoid of parallelisms between hard surfaces, have no sound-focusing concave surfaces, be free of vibrations from external and internal sources, etc. It is for the purpose of providing first design principles of sound recording studios that this book has been prepared, so that for any given size structure, satisfactory vocal and instrumental recording conditions can be established therein.

All equations involving physical quantities are given both in the English and the MKS system of measurement. Also, when the description of existing studios includes linear dimensions, their metric equivalents follow in parenthesis, as is also done for such quantities as surface density (mass per unit area) and sound absorption.

CONTENTS

PART I

BASICS

Chapter 1

Introduction

1.1 GENERAL

The word studio derives from the Latin *studium*, denoting an artist's working place—the artist being a painter, musician, sculptor, dancer, etc.

In modern usage, an explanatory noun is generally prefixed to the word, e.g., film, motion picture, television, and radio studios. However, a studio may also designate a complex of buildings, including offices, laboratories, theaters, etc. as in Universal Studios, Paramount Studios, etc.

The design of a film or television studio is usually a team effort. It may include the production and engineering personnel of the enterprise, an architect, mechanical, structural, and electrical engineer, a lighting and acoustical consultant, an interior designer, and other experts.

Always, nowadays, an environmental impact report must first be prepared before much planning can be done. From the noise point of view, the investigation must consider not only the intrusion of acoustic disturbances from the existing environment, the so-called noise-immission effect, but also the effects which the new enterprise may have on the surroundings, particularly when they include residential areas. Thus the air-conditioning system equipment in the studio, transformers, truck traffic, pumps, and other sources must be evaluated as to their influence on the privacy of the neighborhood.

1.2 DEFINITIONS OF TERMS

The description of the following technical terms employed in architectural acoustics and in the field of noise control are in conformity with those contained in standards by the American National Standards Institute, the American Society for Testing Materials, the Environmental Protection Agency, the California Administrative Code, and the California Bureau of

Aeronautics.

Noise level measurements generally refer to A-weighted sound levels. This is a single-number index for noise containing a wide range of frequencies, and is intended to represent the subjective impression received of the noise by the ear. Since the human hearing mechanism does not respond equally well to sounds of all frequencies, but is less efficient for the lower notes than for the treble, allowance is made in the acoustic data acquisition system to reduce or "weigh" the aural influence of the base with respect to the high-frequency components of the acoustic disturbance under observation. The electronic device employed in the standard sound level meter for this purpose is designated the "A" network, and sound levels so read or recorded carry the subscript A, as in 56 dB-A.

A differentiation should be made between a noise level measurement and a noise exposure level measurement. The former represents no more than the reading by the eye of a hand-held sound level meter at a given place and time. The noise exposure level measurement requires the assessment of the durations of the various noise events which constitute the noise history at the location. By instrument or computations the acoustic data is used for the statistical evaluation of the noise climate. Thus, L_{10} is the decile A-weighted noise level exceeded for 10 percent of the test period.

A differentiation should also be made between ambient noise and exceeding or extraordinary noise. Municipal antinoise ordinances concern themselves chiefly with "home-made" or exceeding noise like that generated by air-conditioning system equipments about a building, the music levels transmitted from a distant amphitheater, lawn mowers, etc. They do not affect noise level limits preempted by the state, as in the case of automobile din (86 dB-A at 50 ft in California) or the federal government, as in the case of interstate (not intrastate) aircraft. Hence, to learn of the ambient noise level at a specific place, the noise must be measured between car passbys, so that only the all-encompassing noise associated with the environment is assessed. This is usually a composite of sounds from many sources, near and far.

In the forefront of statistical terms for the determination of an acoustic climate is the noise equivalent level L_e. It is that level of a hypothetical steady sound, with the same spectrum as the noise under consideration, which has the same total energy as the actual time-varying acoustic disturbances, that is, the noise history at a given place with its temporal sound amplitude variations.

EPA and many municipal planning and building departments prefer a term abbreviated as L_{dn}, the equivalent day-night sound level. It employs the L_e's of 15 daylight hours and the L_e's of 9 night-time hours to which a 10 decibel penalty is affixed.

To measure the sound insulative quality of a sound barrier, several terms have come to be used to express this quantity. Foremost of these is the sound transmission loss, *TL*, or the sound transmission loss characteristic, the variation of the *TL* with frequency.

By definition, *TL* represents the ratio, expressed in decibels, of the sound power incident on a partition to the sound power transmitted through it and radiated by the partition on the far side.

Another term used to express the sound insulative quality of a wall or partition is the *STC* (Sound Transmission Class) rating. It is obtained by a special procedure developed by the American Society of Testing and Materials, and is applicable only to interior partitions. It is a somewhat "puffed up" single-number index which is meant to express the average sound insulation, although it disregards any superior low-frequency insulation of materials. Also, important information is hidden in fine print (like the above-mentioned inapplicability of this rating system to exterior building boundaries).

The reverberation time of a room is defined as the time required for the mean-square sound pressure level, originally in a steady state, to decrease 60 decibels after the source is stopped. The variation of the reverberation period with frequency is known as the reverberation time characteristic.

A decibel denotes the ratio of two quantities in logarithmic form. The denominator of the ratio is always a reference quantity, like 0.0002 microbars for the measurements of sound pressure, or 1 picowatt in the case of sound power. The decibel is often likened to the smallest loudness level change detectable by the human ear.

Chapter 2

Studio Planning

To learn as much as possible about the required sound insulation of the walls and roof of a studio planned to be constructed at a given site, it is desirable to conduct a 24-h noise exposure level survey there. The word "exposure" implies the statistical evaluation of the noise levels at the site and their durations. Actually, for a complete description of the acoustic disturbances it would be necessary to provide a three-dimensional model of the noise spectrum, that is, frequency, level, and the variation of these two quantities with time. But this is an awkward method of noise evaluation, and for most purposes not necessary. What the investigator wants to know chiefly is the maximum noise levels against which the buildings should be shielded, together with the prevailing noise spectrum.

Fig. 2.1 shows the results of a 14-h noise exposure level survey conducted at a proposed studio site. Note that the noise levels are evaluated in terms of the C-weighted sound levels, and thus approach the sound pressure levels. The usual A-weighted sound levels employed in community noise evaluations are of less service here, because they are intended to provide a measure of the annoyance existing in the region. The C-weighted sound levels provide information about the sound insulation of the exterior boundaries of the planned structure, because it is the difference of the external sound pressure levels at different octaves and the required internal sound pressure levels at the same octaves which describes the required sound transmission loss characteristic of the walls and roof of the studio or stage.

Other factors which should be evaluated during the planning sessions concern the following:

1. For a TV studio, should one or two video control rooms be employed, together with an audio control room, and should they be in-line or can, say, the color control room, be located elsewhere? Should any of the control rooms have a window overlooking the

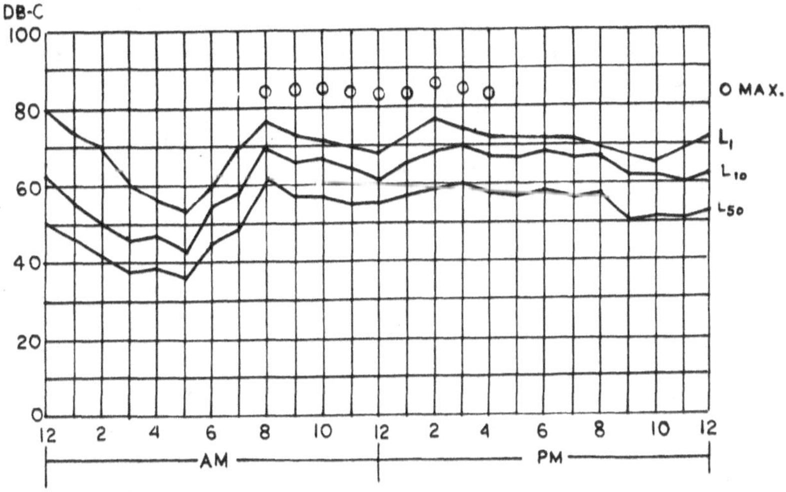

Fig. 2.1 24-h noise exposure level survey at a planned studio site. L_1, L_{10}, and L_{50} represent percentile noise levels exceeded respectively 1, 10, and 50 percent of the observation period. Circles are aircraft flyover noises.

stage, or should they all be "blind"? Is front-screen or rear-screen projection desirable for the stage? Are rain scenes contemplated, so that floor drains should be incorporated in the building? If smoke scenes are required for production, a separate ventilating system besides the air-conditioning one will be necessary. Should the cyclorama screen be recessed in the floor? Should the stage have a camera pit in the floor, or a pool?

2. To achieve very "clean" sound tracks in multiple track recording of music, the signals from the various band sections should not enter the microphone of adjoining sections. To obtain the best recording for such an orchestra, should the different instruments be housed in separate enclosures, or should they be recorded sequentially?
3. Should "Foley Pits" be incorporated in so-called looping or automatic-dialogue replacement chambers, so that sound effects can be generated in the enclosure?
4. Should reverberation chambers be built or should reverberatory effects be obtained with electronic reverberators?
5. Should a portable music shell be considered along with a TV or cinema stage?
6. If electronic lighting is planned to be used in a stage, where should the lighting control panel be located?

7. Besides the direct production facilities there are also pre- and post-production quarters required in many studios. The former may employ vocal or instrumental recording enclosures, with the programs played back during direct production, the performers lip-synching the signals reproduced to them over a loudspeaker or small earphones. Post-production enclosures involve generation of an air or screen master, a composite tape or film which is properly edited, color-corrected, complete with sound effects, etc. Additional rooms may be devoted to the construction of miniature sets to be used in such a system as Magicam, an electronic servo control system coordinating the movements of two cameras, one of which records a foreground action against a blue screen while the other records a miniature set intended to serve as a background for the live action foreground. The foreground camera may be either a film or a video unit. The background camera, which may also be a film or video unit, photographs the miniature set while the electronic servo device slaves the background camera to the foreground or master camera.

Chapter 3

Sound Insulation

The sound insulation capability of a barrier is expressed as its transmission loss, *TL*, at various frequences. *TL* refers to the ratio, expressed in decibels, of the sound power incident on the wall or partition to the sound power transmitted through the barrier and radiated by it. In the laboratory and in the field, when one measure this *TL*, one generally determines the difference in the sound pressure levels on the two sides of the barrier, and then "normalizes" for the acoustics of the receiving room where the transmitted signal is measured.

Barriers of any kind have numerous resonance frequencies at which the *TL* becomes rather small. In the lengthy frequency region between the very high and low resonances, where the barrier is chiefly "mass controlled," its insulation is dependent mainly on its surface density, that is, its mass per unit area.

The equations expressing the acoustic mass law are

$$TL = 14.5 \log M + 20 \log f - 31$$

$$TL = 14.5 \log M' + 20 \log f - 41$$

where

M = surface density of barrier in lb/ft^2
M' = surface density of barrier in kg/m^2

Fig. 3.1 shows the variation of the *TL* of a monolithic, nonporous barrier with frequency for different surface densities. For each doubling of M or M' the *TL* increases 4.35 dB at all frequencies, and for a quintupling of M or M' by 10 dB.

The figure indicates that for high *TL*'s, particularly at the low notes, the surface density of a wall or roof becomes impractically large. For instance, to achieve a *TL* of 48 dB at 125 hertz, the wall or roof density would have to be 320 lb/ft^2 or 1562 kg/m^2.

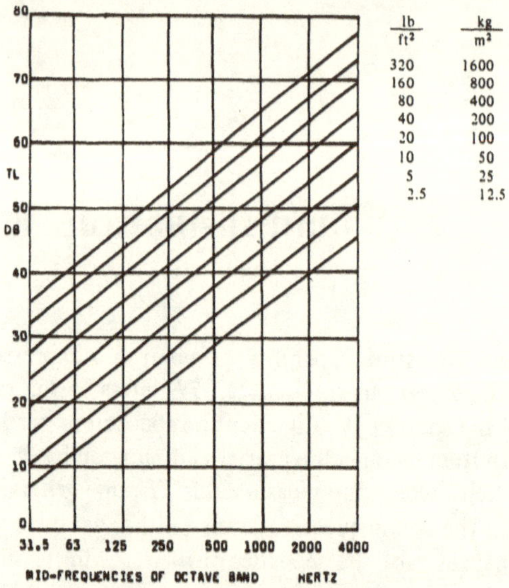

Fig. 3.1 Variation of sound transmission loss with frequency of a monolithic barrier for different surface densities.

When high sound insulation at low frequencies is required, a monolithic barrier becomes too thick and heavy, and recourse has to be taken to a double barrier. The two "leaves" of such a construction should be isolated from each other as much as possible, since "sound bridges" between the two tend to lower the overall transmission loss of the barrier. This may mean separate foundations for the two walls, with a minimum of resilient lateral support between the two, and a generous amount of sound-absorbent material in the interspace. The purpose of the sound-absorbent is to reduce the effect of the standing waves between the panels in both the horizontal and the vertical dimensions. Hence the absorbent should be placed not only on the walls themselves, but also around the periphery of the cavity.

For a double barrier with monolithic, nonporous panels, the empiric mass law becomes

$$\begin{aligned} TL &= 20 \log M + 13 \log D + 20 \log f - 39 \\ &= 20 \log M' + 13 \log D' + 20 \log f - 32 \end{aligned}$$

where

M = surface density of double barrier in lb/ft^2
M' = surface density of double barrier in kg/m^2
D = distance between barrier in inches
D' = distance between barriers in meters
f = frequency.

The relationship is illustrated graphically in Fig. 3.2 for $f = 125$ hertz, since this is usually the critical frequency, demanding the highest surface density of the construction. For this frequency, the equation for the transmission loss becomes

$$TL = 20 \log M + 13 \log D + 3$$
$$= 20 \log M' + 13 \log D' + 10$$

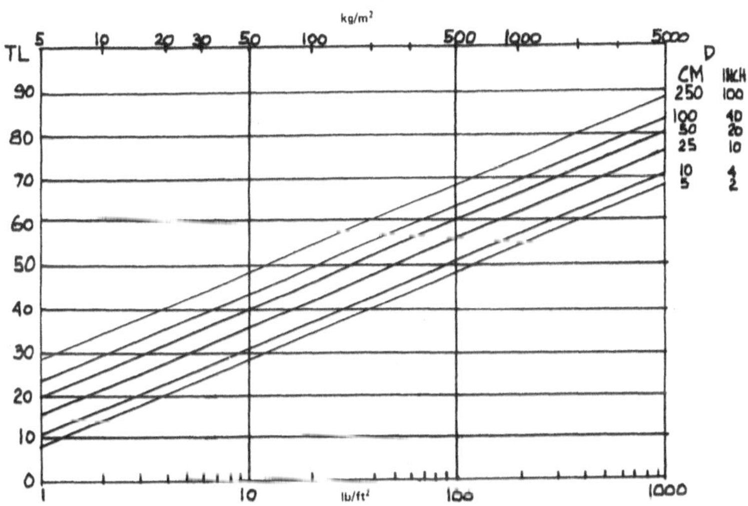

Fig. 3.2 Variation of 125-Hz sound transmission loss with surface density of a double barrier for various distances of separation D between the two panels.

Fig. 3.3 gives a graphic example for the calculation of the insulation of the exterior boundaries of a planned studio exposed to a noise level characteristic shown by the solid line marked Curve A. It corresponds to an A-weighted sound level of 75 dB-A measured at the boundary of the builid-ing. Curve B represents the desired interior noise level spectrum for the room, and comes to PNC-20 or 30 dB-A. The differences between Curve A and B at the various frequencies are depicted by the dot-dash line labeled Curve C.

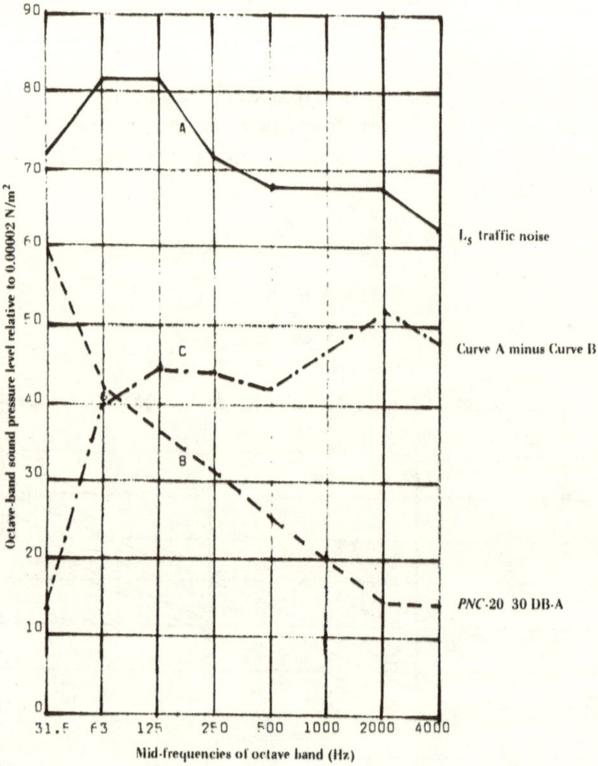

Fig. 3.3 Curve A: Street traffic noise exceeded 5 percent of time; Curve B: Required inside noise level characteristic; Curve C: Necessary sound insulation characteristic of stage walls.

From Figs. 3.1 and 3.2 it can readily be learned that, at 125 hertz, when the required *TL* comes to 45 dB, a single monolithic wall would have to have a surface density of nearly 200 lb/ft^2 (976 kg/m^2), comparable to almost 1.5 ft (.46 m) of solid concrete. A double wall, on the other hand, depending on the distance of separation between the two leaves, can be remarkably light, as shown below for walls with a *TL* of 45 dB at 125 hertz having separation distances and surface densities shown below:

D'	D	M'	M
5 cm	2 in.	400 kg/m^2	80 lb/ft^2
10 cm	4 in.	250 kg/m^2	50 lb/ft^2
15 cm	6 in.	200 kg/m^2	40 lb/ft^2
20 cm	8 in.	160 kg/m^2	32 lb/ft^2
25 cm	10 in.	140 kg/m^2	28 lb/ft^2
61 cm	24 in.	80 kg/m^2	16 lb/ft^2

It must be realized that for small distances of separation it is practically impossible to achieve the desired *TL*'s given by the equation and the above table, unless extraordinary precautions are taken. For instance, while a 2 in. (5 cm) separation between the two glass panes of a monitoring room window can generally be readily accomplished with only a perimeter clamp, structural walls must have lateral elastic restraints. In such cases it is practical to add, for the *TL* of a double wall of surface density M, 6 dB to the *TL* of a single wall of equal M.

It is a relatively simple matter to calculate the required sound insulation for a gabled roof for which there is no ceiling. This can be done by calculating the necessary surface density of the roof from a knowledge of the existing external and the desired internal noise level characteristics.

When there exists a ceiling below a roof, the required sound insulation must be found in published information about such constructions.

When a building has a flat roof and no ceiling below it, as is the case of many sound stages, and the building is exposed to noise from only one direction, the required roof insulation must consider the diffracted sound on the roof when it is not subject also to direct sound from the sources. Fig. 3.4 shows the excess sound level attenuation (over the direct sound level) as a function of the effective barrier height H for various diffraction angles when the significant frequency is 125 hertz. A line from the source S is drawn through a point on the roof representing half the length of the building. From the geometric construction of the situation, the diffraction angle θ is determined. As an example, when the stage is 100 ft (30 m) long and the source is 150 ft (45 m) from a building, that is 40 ft (15 m) high, the diffraction angle comes to 14 degrees and the barrier height to 12.5 ft (3.6 m). In this case, the excess sound attenuation at 125 hertz is 11 dB.

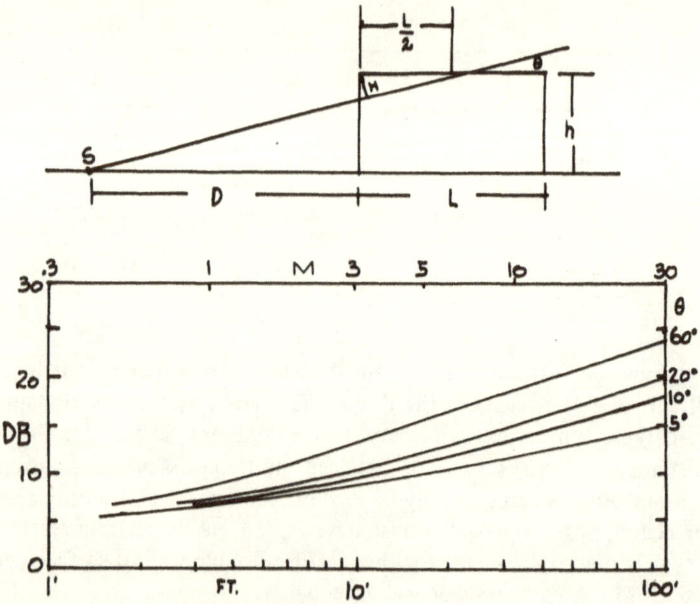

Fig. 3.4 Variation of 125-Hz excess sound attenuation with effective barrier height H for various angles of diffraction. For the excess sound attenuation at higher octaves, add 3 dB per octave.

The equation for the barrier height is

$$H = (h^2 + D^2)^{1/2} \sin (\tan^{-1} (h/D) - \tan^{-1} [h/(D + 0.5L)]$$

$$\cong h \left(1 - \frac{1}{1 + \frac{L}{D}} \right)$$

For buildings of low height, a parapet on the roof will increase the effective barrier height H. The angle θ is

$$\theta = \tan^{-1} \frac{h}{D + \frac{L}{2}}$$

In calculating the noise level in an enclosure whose walls and roof are subjected to different noise levels, consider a 20 ft × 37 ft × 50 ft (6 m × 11 m × 15 m) sound stage with a total sound absorption of 4000 sabins (372 metric sabins) at 125 Hz. The concrete floor is assumed to be negligibly sound transmissive so that its *TL* does not need to be considered in the calculations. Let the 740 ft^2 (68 m^2) west wall have a 125-Hz *TL* of 45 dB, and be subjected to a sound level of 82 dB at 125 Hz. The areas of the south, north, and east walls add up to 2740 ft^2 (255 m^2) with a *TL* of 43 dB at 125 Hz, and are exposed to a sound level of 76 dB at 125 Hz. The 1850 ft^2 (172 m^2) roof with its *TL* of 40 dB at 125 Hz is impacted by a sound level of 70 dB at 125 Hz. What is the total noise level in the enclosure?

The various sound level reductions are as follows:

$$SLR_w = 45 - 10 \log \frac{740}{4000} = 52.3 \text{ dB}$$

$$SLR_{SNE} = 43 - 10 \log \frac{2740}{4000} = 44.6 \text{ dB}$$

$$SLR_R = 40 - 10 \log \frac{1850}{4000} = 43.3 \text{ dB}$$

The 125-Hz sound levels in the enclosure produced by the different boundaries are:

$$SL_W = 82 - 52.3 = 29.7 \text{ dB}$$

$$SL_{SNE} = 76 - 44.6 = 31.4 \text{ dB}$$

$$SL_R = 70 - 43.3 = 26.7 \text{ dB}$$

The logarithmic sum of the 3 sound levels comes to

$$SL = 10 \log (10^{29.7/10} + 10^{31.4/10} + 10^{26.7/10}) = 34.4 \text{ dB}$$

By inspecting the three terms in the parenthesis of the above equation it is seen that the center one is by far the largest. It contributes significantly to the resulting noise level in the stage, while the noise level transmitted through the roof is the least.

The general equation for the sound level in a stage with boundary surfaces $S_1, S_2, \ldots S_n$, with corresponding sound transmission losses TL_1, $TL_2, \ldots TL_n$, corresponding impacting noise levels $NL_1, NL_2, \ldots NL_n$, while the stage has a total sound absorption A, is given by

$$SL = -10 \log A + 10 \log (S_1 \times 10^{0.1 \Delta L_1} + S_2 \times 10^{0.1 \Delta L_2} + \ldots S_n \times 10^{0.1 \Delta L_n})$$

where

$$\Delta L_i = NL_i - TL_i$$

When every one of the n boundaries of a building transmits the same amount of noise energy towards a total interior sound level SL, the equation becomes

$$\frac{10^{SL/10}}{n} = \frac{S_i \times 10^{(NL_i - TL_i)/10}}{A}$$

$$TL_i = NL_i - SL + 10 \log \frac{nS_i}{A}$$

Thus, in regard to the previous example, if it had been desired to have an internal noise level of 30 dB, and the 740-ft² (68 m²) west wall was faced with a noise level NL = 82 dB at 125 Hz, the TL of that wall at 125 Hz would have had to be

$$TL_W = 82 - 30 + 10 \log \frac{3 \times 740}{4000}$$

$$= 49 \text{ dB}$$

Concrete blocks are frequently employed in the construction of cinema and television stages because of their high surface density and reasonable cost. Where lead costs 70¢/lb ($1.54/kg), concrete may be had at a small fraction of that price per pound (1¢/lb).

Concrete blocks are listed in three categories—hollow loadbearing, hollow non-loadbearing, and solid loadbearing. A solid concrete block is defined as a unit in which the core area is not more than 25 percent of the gross cross-sectional area. A hollow concrete block is a unit having a core area greater than 25 percent of the gross cross-sectional area. Generally, the core area of a hollow unit is from 40–50 percent of the gross area.

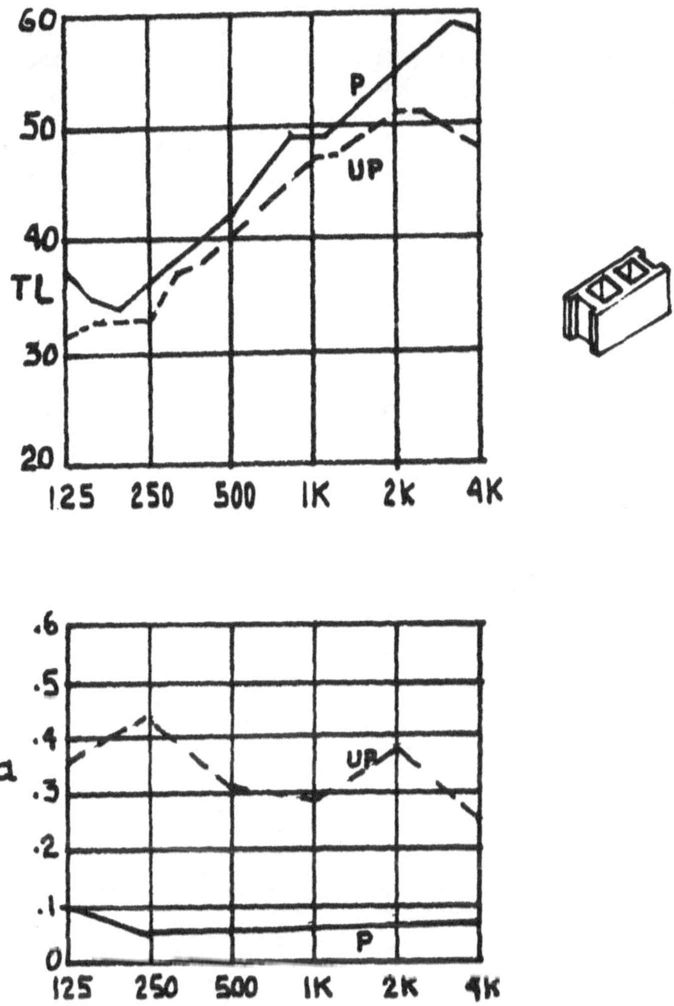

Fig. 3.5 Top diagram shows the sound transmission loss characteristic of a painted (P) and an unpainted (UP) 8 in. (20 cm) thick hollow-core concrete block wall with vertical mortar joints. Bottom diagram shows the sound absorptivity characteristic of such a concrete block wall, painted and unpainted.

Fig. 3.5 shows the sound transmission loss characteristic of a wall built with 8 in. (20 cm) hollow-core concrete blocks with vertical mortar joints, when the wall is unpainted and when it is painted. The lower part of the figure gives the reason for the higher sound transmission loss of the painted concrete block wall—the unpainted wall has a high sound absorption coefficient at all frequencies within the audio spectrum, compared to the painted wall. Since nearly five times as much sound energy can enter the unpainted surface, it stands to reason that considerably more sound energy can pass through the block and become radiated on the other side of the unit.

For maximum sound insulation of a concrete block wall it is recommended that all cores of the block be filled with concrete during the erection of the wall, not only to increase its surface density, but also to block all apertures inside the cores through which sound may become transmitted.

The paint used for closing the external pores of a concrete block is not important, as long as it can make good contact with the material and will not later flake off.

From the upper diagram of the figure it is seen that the sound transmission loss of the painted concrete block wall is close to 3 dB greater than for the unpainted one. This means that the paint has reduced the transmission of the sound energy through the material by 50 percent. This is equivalent to another unpainted concrete block wall which is 60 percent thicker, that is, to a wall made of 12 in. (30 cm) thick instead of 8 in. (20 cm) thick blocks, according to the acoustic mass law.

The table below lists four building materials, each of such thickness that 1 ft^2 weighs 4 lb (1 m^2 weighs 19.5 kg), thus providing a sound transmission loss of 32 dB at 500 Hz.

	Thickness		*Price*		*Price*	
Material	in.	cm	per lb	per kg	per ft^2	per m^2
Sheet lead	1/16	0.16	$0.70	$1.54	$2.80	$30.00
Plywood	1	2.54	0.20	0.44	0.80	8.56
Plasterboard	1	2.54	0.08	0.17	0.32	3.42
Concrete	.38	0.95	0.01	0.02	0.04	0.43

The above prices do not include the cost of installation, which is different for each material. It was prepared to show that, pound for pound, concrete is by far the cheapest building material for a given sound insulation.

Solid concrete walls are made by pouring the material into forms for these boundaries, or the walls may be of the tilt-up type, in which the boundary is poured on the ground, after which the panel is raised by a crane to a vertical position. Steel H- columns hold the panels in place laterally while they are resting on a concrete foundation cast before the panels are cast.

In quiet areas it is possible to build sound and TV stages by the wood-frame construction method. A double-stud wall is preferred over a staggered-stud one, for increased sound insulation. The double-stud wall rests on two plates separated from each other by at least 2 in. (5 cm), while the staggered-stud type employs only one top and bottom plate. In the latter type of building it is not possible to employ so-called horizontal fire-stops for bracing between the studs, because the opposite row of studs intrudes on the other row. This reduces the stiffness of the boundary, with the result that the wall may exhibit pronounced resonance dips in the low-frequency region.

Fig. 3.6 shows the sound transmission loss of various wood-frame constructions.

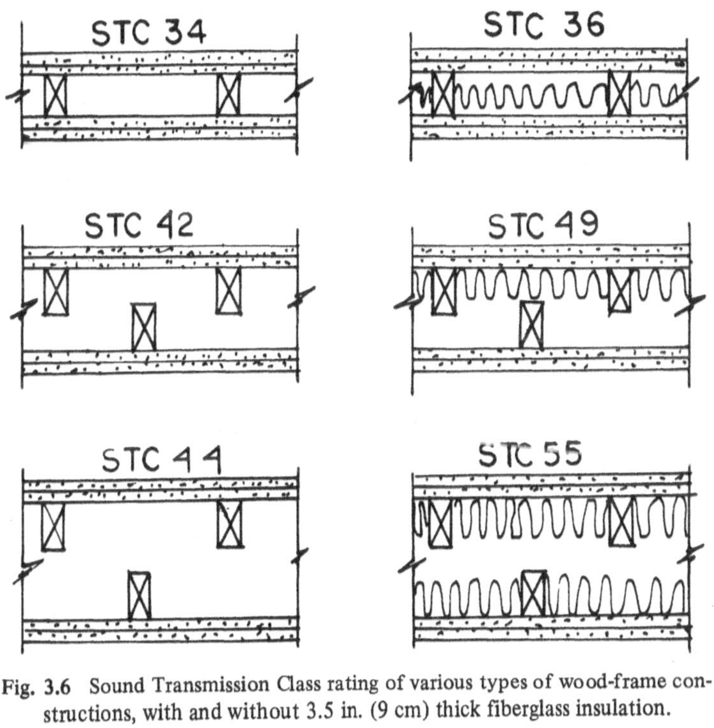

Fig. 3.6 Sound Transmission Class rating of various types of wood-frame constructions, with and without 3.5 in. (9 cm) thick fiberglass insulation.

Chapter 4

Reverberation

4.1 GENERAL

Every enclosed space and most partially enclosed spaces (e.g., tunnels) exhibit the phenomenon of reverberation. It is defined as the persistence of sound due to multiple reflections in a room after the source has stopped. This prolonged audibility of the signal tends to reduce the intelligibility of the spoken word by causing overlapping of successive syllables. For music, a slightly prolonged signal persistence in the room is generally found to add to the quality of the renditions.

Over the decades, two reverberation time equations have found favor among the acousticians, namely, the Sabine and the Eyring-Norris formulas. The former is generally employed in moderately live enclosures, with an average absorptivity of less than 0.20, while the latter is used in predicting the period in "dead" rooms.

The Sabine and Eyring-Norris equations are:

English Units	Metric Units
$T_s = \dfrac{0.05V}{A + 4mV}$	$= \dfrac{0.16V'}{A' + 4m'V'}$
$= \dfrac{0.05V}{\Sigma S'_i a_i + 4\,mV}$	$= \dfrac{0.163V'}{\Sigma S'_i a_i + 4m'V'}$
$= \dfrac{0.05V}{S\bar{a} + 4mV}$	$= \dfrac{0.163V'}{S'\bar{a} + 4m'V'}$

$$T_E = \frac{0.05V}{-S\lg(1-\bar{a}) + 4mV} = \frac{0.163V'}{-S'\lg(1-\bar{a}) + 4m'V'}$$

$$= \frac{0.05V}{-2.3\log_{10}(1-\bar{a}) + 4mV} = \frac{0.163V'}{-2.3\log_{10}(1-\bar{a}) + 4m'V'}$$

where

V = volume of room in ft^3
V' = volume of room in m^3
S = surface of room in ft^2
S' = surface of room in m^2
A = absorption in room in sabins
A' = absorption in room in metric sabins or m^2
$\bar{a}$ = average absorptivity of room boundary material
m = molecular absorption of sound in air in 1/ft
m' = molecular absorption of sound in air in 1/m
$\bar{a}$ = average absorptivity

$$= \frac{\Sigma S_i a_i}{S} = \frac{\Sigma S_i' a_i}{S'}$$

The table below gives $4m$ and $4m'$ for various frequencies, relative humidities, and temperatures:

Relative humidity	*Temperature* C F	2000 Hz		4000 Hz		63000 Hz	
		$4m$	$4m'$	$4m$	$4m'$	$4m$	$4m'$
30%	15 (59)	.0044	.0143	.0148	.0486	.0322	.01056
	20 (68)	.0036	.0119	.0116	.0379	.0256	.0840
	25 (77)	.0035	.1114	.0095	.0313	.0209	.0685
	30 (86)	.0034	.0111	.0086	.0281	.0172	.0564
50%	15 (59)	.0030	.0099	.0087	.0286	.0191	.0626
	20 (68)	.0029	.0096	.0074	.0244	.0153	.0503
	25 (77)	.0029	.0095	.0072	.0235	.0135	.0444
	30 (86)	.0028	.0092	.0071	.0233	.0130	.0426
70%	15 (59)	.0027	.0088	.0068	.0223	.0138	.0454
	20 (68)	.0026	.0085	.0065	.0213	.0122	.0399
	25 (77)	.0026	.0084	.0064	.0211	.0118	.0388
	30 (86)	.0025	.0082	.0063	.0207	.0117	.0383

The reverberation time of a room is defined as the time required for the mean-square sound pressure level therein, originally in a steady state, to decrease 60 dB after the source is stopped. Unless otherwise noted, it refers to the standard frequency of 1000 Hz.

The variation of the reverberation time with frequency is known as the reverberation time characteristic.

In reverberation time calculations for rooms with a volume less than 5000 ft^3 (141 m^3), the air absorption for the treble is rarely included because it represents no more than 5 percent of the boundary absorption, while for an enclosure as large as 10^6 ft^3 it can come to as much as 50 percent, depending on the frequency, temperature, and relative humidity in the room.

It can be proven that for a frequency of 10,000 Hz, at 68 F and relative humidity $\leqslant$ 35 percent, no room can have a reverberation time of more than 1 s, regardless of volume or shape. Thus, if we assume the boundary absorption to be zero, we have for the reverberation time $T = 0.05/4m$; so that when $4m$ comes to 0.05 or larger, T becomes $\leqslant$ 1 s.

Since dry air is more absorbent than moist air it has become practice in recording studios to call for a relative humidity of not less than 50 percent, to preserve the high notes as much as possible in the room.

To prolong the treble reverberation period it is also desirable to employ plywood panels for the wall treatment, since they tend to be well absorptive for the lows but relatively reflective for the higher frequencies.

It may be desirable at this point to consider the Eyring-Norris equation a little more closely, and without the term representing the molecular absorption, that is

$$T_E = \frac{0.05V}{-S\lg(1-\bar{a})} = \frac{0.163V'}{-S'\lg(1-\bar{a})}$$

$$= \frac{0.05V}{-2.3S\log(1-\bar{a})} = \frac{0.163V'}{-2.3\,S'\log(1-\bar{a})}$$

where

lg = logarithm to the base e
log = logarithm to the base 10
V = room volume (ft^3)
V' = room volume (m^3)

S = total interior surface of the room (ft^2)
S' = total interior surface of the room (m^2)

The denominators in the above equations represent the total sound absorption required in the enclosure to achieve a reverberation time T. Fig. 4.1 shows the absorption in sabins as a function of the room volume in cubic feet for various reverberation times. Fig. 4.1A shows this variation in the metric system.

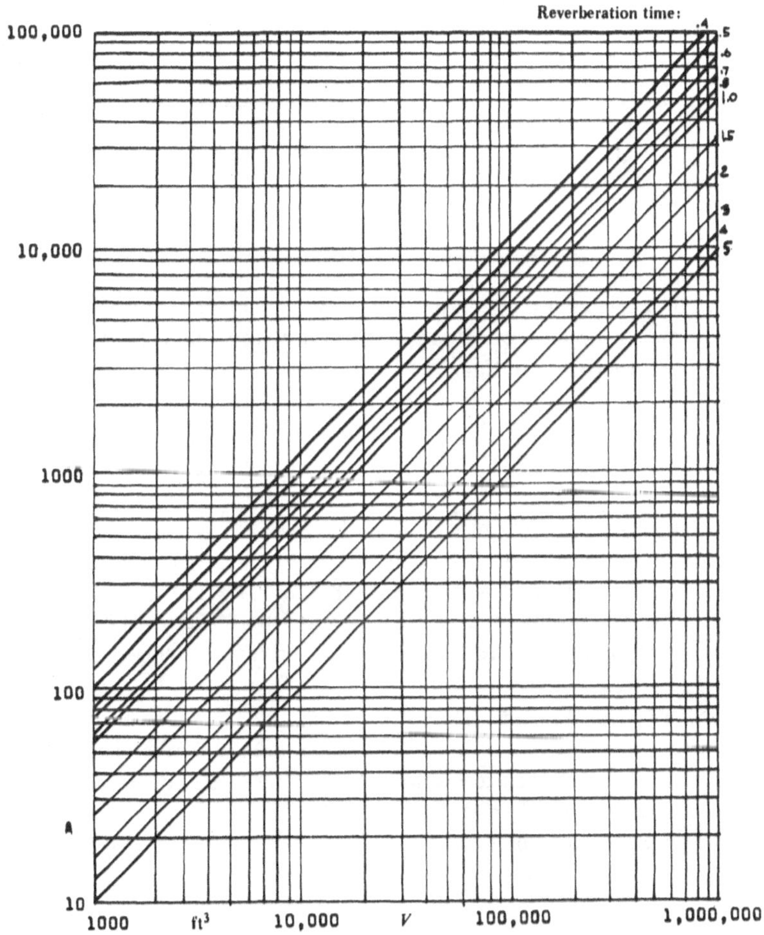

Fig. 4.1 Variation of sound absorption in sabins with room volume in cubic feet, for various reverberation times in seconds.

Fig. 4.2 shows a table which provides the value of $-\lg(1-\bar{a})$ for various values of $\bar{a}$, the average absorptivity of the room's boundary material.

To secure a graphic representation of $\bar{a}$ as a function of the room volume, the problem of finding an approximate expression for the ratio V/S exists. This turns out to be $0.15V^{1/3}$ in both the English and the metric systems. As an example consider various shaped rooms with a volume of 6000 ft^3 (170 m^3). The table below contains the significant information for the approximation $0.15V^{1/3} = 0.15 \times 6000^{1/3} = 2.8$):

Basic Ratio	*Dimensional Ratio*	*V/S*
1:2:3	10 ft × 20 ft × 30 ft	2.73
2:3:5	10.7 ft × 17.5 ft × 29.2 ft	2.82
3:5:8	11.1 ft × 18.4 ft × 29.5 ft	2.81
	average	2.79

By the Eyring-Norris equation the required average sound absorptivity of the interior boundary material of a room with volume V, interior surface S, and reverberation time T is

$$\bar{a} = 1 - \frac{1}{10^{0.05V/2.3ST}} = 1 - \frac{1}{10^{0.163V'/2.3S'T}}$$

By employing the expression $0.15\,V^{1/3} = V/S$, one may plot $\bar{a}$ as a function of V for various T's, as shown on Fig. 4.2. Thus, for a 50 ft × 125 ft × 160 ft (15.2 m × 38.1 m × 48.8 m) stage with a volume of 10^6 ft^3 (28,300 m^3), a total interior surface of 6.8×10^4 ft^2 (6317 m^2), and a desired reverberation time of 0.7 s at 500 Hz, the average absorptivity comes to 0.65.

However, this is not the absorptivity of the wall and roof material when the floor is hard. For such a stage, with its 2.0×10^4 ft^2 (1858 m^2) of concrete floor, the average boundary absorptivity is more nearly $(6.8 \times 10^4)(0.65) / 4.8 \times 10^4) = 0.92$.

Substituting the expression $V/S = 0.15V^{1/3}$ in the above equation for $\bar{a}$ yields

$$\bar{a} = 1 - \frac{1}{10^{0.0032V^{1/3}/T}} = 1 - \frac{1}{10^{0.01V'^{1/3}/T}}$$

$\bar{a}$	$-\log_e(1-\bar{a})$	$\bar{a}$	$-\log_e(1-\bar{a})$	$\bar{a}$	$-\log_e(1-\bar{a})$	$\bar{a}$	$-\log_e(1-\bar{a})$
.005	.0050	.255	.2944	.505	.7032	.755	1.4065
.010	.0101	.260	.3011	.510	.7133	.760	1.4271
.015	.0151	.265	.3079	.515	.7236	.765	1.4482
.020	.0202	.270	.3147	.520	.7340	.770	1.4697
.025	.0253	.275	.3216	.525	.7444	.775	1.4917
.030	.0305	.280	.3285	.530	.7550	.780	1.5141
.035	.0356	.285	.3355	.535	.7657	.785	1.5371
.040	.0408	.290	.3425	.540	.7765	.790	1.5606
.045	.0460	.295	.3496	.545	.7875	.795	1.5847
.050	.0513	.300	.3567	.550	.7985	.800	1.6094
.055	.0566	.305	.3638	.555	.8097	.805	1.6348
.060	.0619	.310	.3711	.560	.8210	.810	1.6607
.065	.0672	.315	.3783	.565	.8324	.815	1.6874
.070	.0726	.320	.3857	.570	.8440	.820	1.7148
.075	.0780	.325	.3930	.575	.8557	.825	1.7430
.080	.0834	.330	.4005	.580	.8675	.830	1.7720
.085	.0888	.335	.4080	.585	.8795	.835	1.8018
.090	.0943	.340	.4155	.590	.8916	.840	1.8326
.095	.0998	.345	.4231	.595	.9039	.845	1.8643
.100	.1054	.350	.4308	.600	.9163	.850	1.8971
.105	.1109	.355	.4385	.605	.9284	.855	1.9310
.110	.1166	.360	.4463	.610	.9416	.860	1.9661
.115	.1222	.365	.4541	.615	.9545	.865	2.0025
.120	.1278	.370	.4626	.620	.9676	.870	2.0402
.125	.1335	.375	.4700	.625	.9808	.875	2.0794
.130	.1393	.380	.4780	.630	.9943	.880	2.1203
.135	.1450	.385	.4861	.635	1.0079	.885	2.1628
.140	.1508	.390	.4943	.640	1.0217	.890	2.2073
.145	.1567	.395	.5025	.645	1.0356	.895	2.2538
.150	.1625	.400	.5108	.650	1.0498	.900	2.3026
.155	.1684	.405	.5192	.655	1.0642	.905	2.3532
.160	.1744	.410	.5276	.660	1.0788	.910	2.4085
.165	.1803	.415	.5361	.665	1.0936	.915	2.4652
.170	.1863	.420	.5447	.670	1.1087	.920	2.5259
.175	.1924	.425	.5534	.675	1.1239	.925	2.5904
.180	.1985	.430	.5621	.680	1.1394	.930	2.6595
.185	.2046	.435	.5709	.685	1.1552	.935	2.7332
.190	.2107	.440	.5798	.690	1.1712	.940	2.8138
.195	.2169	.445	.5888	.695	1.1874	.945	2.9013
.200	.2231	.450	.5978	.700	1.2040	.950	2.9933
.205	.2294	.455	.6070	.705	1.2208	.955	3.1016
.210	.2357	.460	.6162	.710	1.2379	.960	3.2190
.215	.2421	.465	.6255	.715	1.2553	.965	3.3526
.220	.2485	.470	.6349	.720	1.2730	.970	3.5068
.225	.2549	.475	.6444	.725	1.2910	.975	3.6842
.230	.2614	.480	.6539	.730	1.3093	.980	3.9144
.235	.2679	.485	.6636	.735	1.3280	.985	4.1907
.240	.2744	.490	.6733	.740	1.3471	.990	4.6052
.245	.2810	.495	.6832	.745	1.3665	.995	5.2983
.250	.2877	.500	.6931	.750	1.3863	1.000	

Fig. 4.2 Variation of average absorptivity with room volume for various reverberation times T.

In reality, it may not be possible to achieve a reverberation time of 0.7 s at 500 Hz in a stage so large. The reason is that, although many manufacturers of fiberglass boards 2 in. (5 cm) thick list a sound-absorptivity of 0.99 at 500 Hz for their product, this absorptivity generally refers to an acoustic test sample with an area close to 1/8 the area of the reverberation chamber in which the test was conducted. It has been found, however, that the absorption "coefficient" of an acoustic material is not a constant, but a function of the amount of the material in a room. The larger the surface in a room is covered with the absorbent treatment, the smaller is the absorptivity of the product. Similarly, when only a few square feet of the material are installed in an enclosure, the absorptivity may exceed one. The reason is that a small acoustically treated area in a room represents a low sound-pressure area into which adjoining waves may be drawn in the way a vacuum hose tends to suck in dust particles over a wide area of floor.

Fig. 4.3 shows the variation of $\log_e (1 - \bar{a})$ with $\bar{a}$, the average sound absorptivity. When $\bar{a}$ is small, the $\log_e$ term is approached by $-(\bar{a} + \frac{1}{2}\bar{a}^2)$.

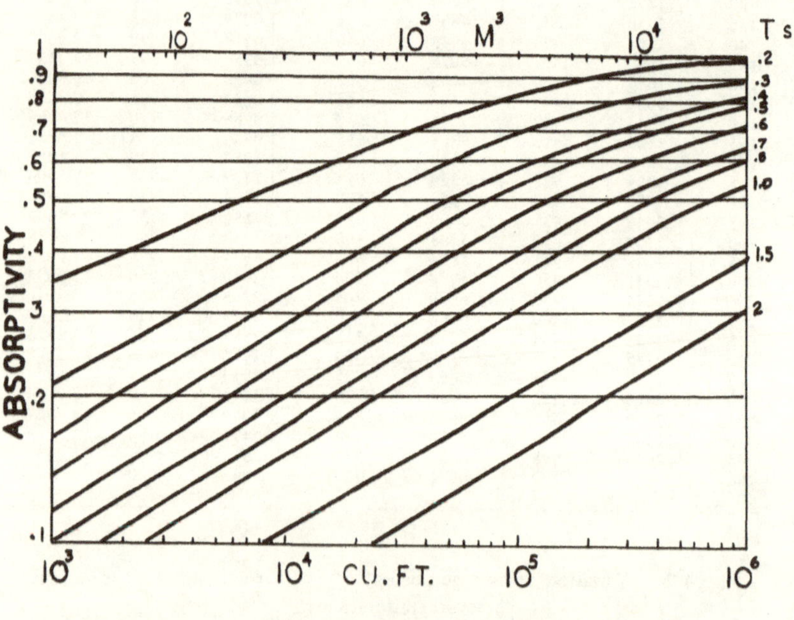

Fig. 4.3 Table giving values of $\bar{a}$, the average absorptivity, and $-\log_e (1 - \bar{a})$

In an auditorium or a theater the optimal reverberation time at a given frequency is usually calculated on the basis that the room is two-thirds occupied. From a knowledge of the absorption per person as a function of frequency it is readily possible to determine the additionally required sound absorption in the enclosure to secure in it the optimal reverberation characteristic.

In sound stages an optimal reverberation time at a given frequency cannot be determined because the room is always kept so dead as to establish outdoor conditions in it. There is, however, the danger that there may exist an undesirably long reverberation period at the low frequencies because of the reduced absorptivity when the acoustic material is applied against a hard backing. Thus, the absorptivity of Owens-Corning No. 703 fiberglass board, 2 in. (5 cm) thick on an AIMA Mounting No. 4 (material laid against the concrete laboratory floor), comes to only 0.22 at 125 Hz, although it measures 0.99 in the frequency range from 500 to 4000 Hz. Because of the low-frequency absorption by the roof structure and the framework of the lighting grid and entrapment of the sound above the grid (simply part of the grid absorption), the reverberation time at 125 Hz measured in a large empty TV stage came to less than the calculated period based entirely on the absorption of the acoustic material applied to the roof and the walls. When such a stage is in operation and there are sets, people, and equipment about the floor, the 125-Hz reverberation time generally measures in the order of 1.4–1.5 s, compared to 0.8–0.85 in the mid-range of the audio spectrum, or 75 percent longer.

Fig. 4.4 shows the absorption characteristics of Owens-Corning No. 703 plain fiberglass board of various thicknesses when the material is mounted on a hard wall, the so-called No. 4 mounting. Note the very large increase in the 250-Hz absorptivity of the 2 in. (5 cm) material against the 1 in. (2.5 cm) one. In large stages the 2 in. board is definitely required because the cyclorama screen's low-frequency absorptivity rarely extends beyond 125 Hz.

In a sound stage used for television, the low-frequency reverberation is further reduced by the vibration of the cyclorama screen which such a room generally contains. This sound absorption will be discussed in greater detail in the chapter on TV and cinema stages.

It is not possible to recommend an optimum 500-Hz reverberation time for a specific enclosure without considering its predominantly functional aspects. This is particularly true for music rooms whose quality of recording or reproduction is judged subjectively to a large degree. As Lord Rayleigh said in the first paragraph of the first chapter of his monumental "The Theory of Sound," all questions connected with the subject of sound must come for a decision to the ear—and from it, there can be no appeal.

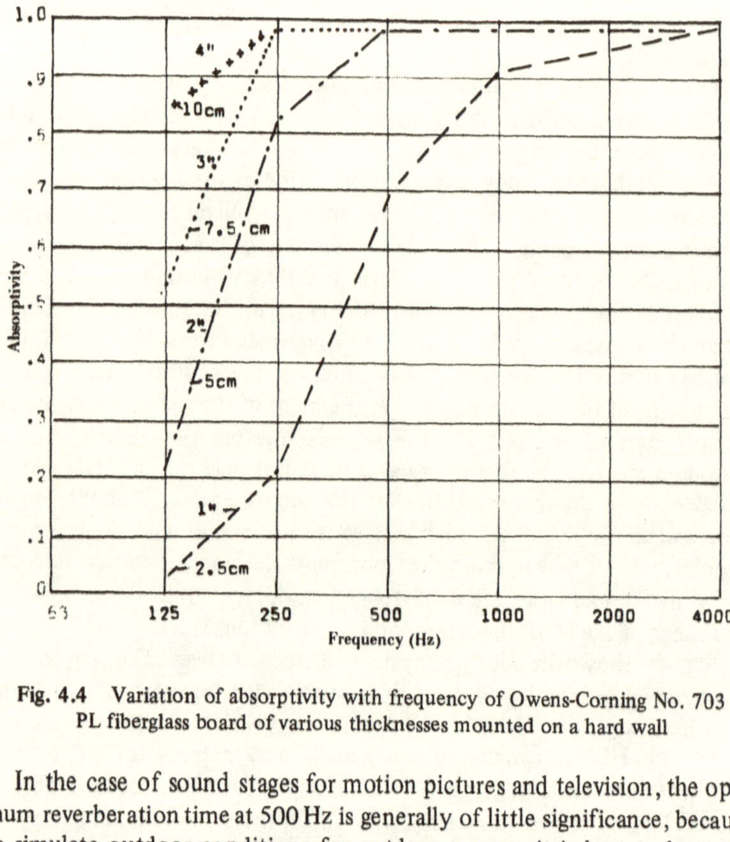

Fig. 4.4 Variation of absorptivity with frequency of Owens-Corning No. 703 PL fiberglass board of various thicknesses mounted on a hard wall

In the case of sound stages for motion pictures and television, the optimum reverberation time at 500 Hz is generally of little significance, because to simulate outdoor conditions for outdoor scenery it is best to keep the reverberation period as low as is practically possible. This does not mean the use of anechoic wedges on all the walls, ceilings, and below an iron grid floor, which treatment would be extravagantly expensive when the enclosure has a volume of $\geqslant 10^6$ ft^3. Nor is it necessary to achieve such a dead stage because the microphone distances are generally so short that reflected sound energy is minimal at the microphone and unnoticeable regardless of the reverberation time existing in the stage.

For music recording studios it becomes important to consider the prevalent use of multiple track recording, which requires a minimum of sound from adjoining instruments to achieve a clean, unmasked recording for a definite band section. Hence, such enclosures are also kept on the dead side.

For these reasons it is recommended to try to achieve a flat reverberation time characteristic rather than a specific period for a specific frequency.

The optimum reverberation time of a recording studio is dependent on several factors, such as the kind of music to be recorded (rock-and-roll, symphonic, religious, etc.), the number of sound tracks to be employed in the recording, and the number of instruments to be recorded. Since most recording studios have to accommodate a wide variety of instruments and music, one can only recommend a mean time or range of reverberation times, depending on the prevalence of the music and the size of the band used in the studio. It is generally best to design the studio towards a low reverberation period, because it is a relatively simple matter to add reverberation electronically, but impossible to reduce this effect electronically.

It is possible to place rotatable panels that are reflective on one side and absorbent on the other, on the sidewalls of a studio in order to vary the reverberation period to some extent—usually by no more than 10 percent. Or one may move an entire sidewall or lower the ceiling, to effect a similar change. However, these means are often highly expensive, and undesirable from other points of view, such as changing the ratio of dimensions.

A satisfactory reverberation period for a recording studio can usually be obtained by employing the equation

$$T = 0.15 \log V$$

$$= 0.15 \log V' + 0.23$$

where

V = room volume (ft^3)

V' = room volume (m^3)

Fig. 4.5 shows the variation of the 500-Hz reverberation time with recording studio volume.

In making reverberation time measurements in a studio it is desirable to employ a sufficiently long distance between the sound source and microphone to achieve consistent results from point to point in the room, as shown in Fig. 4.6. When the distance between the sound source and the microphone is just a few feet, with the microphone placed at a distant corner, the initial period is much shorter than when the microphone is near the middle or the end of the enclosure.

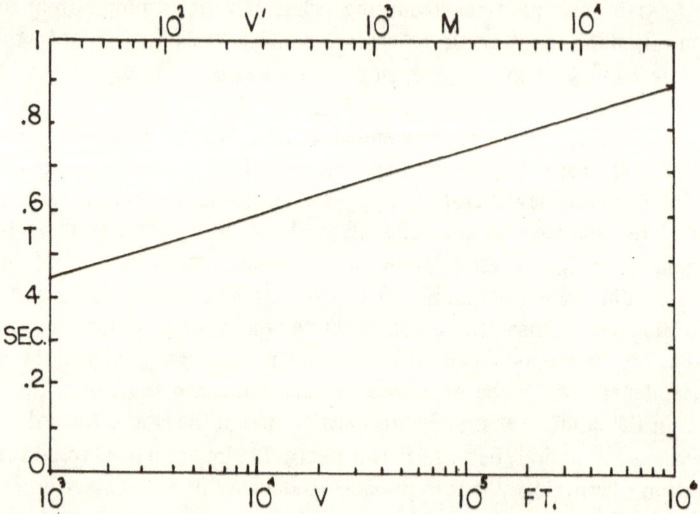

Fig. 4.5 Variation of 500-Hz reverberation time with recording studio volume

The reverberation period is generally evaluated with a Bruhl and Kjar high-speed graphic level recorder, or else digitally with a frequency analyzer and a, say, Hewlett Packard HP 9825 Analyzer.

An Inovonics 500 Spectrium Analyzer has a built-in reverberation time meter and pink noise source, which allow the instant determination of the period.

4.2 RE-RECORDED REVERBERATION

When the reverberation generated in a reverberation chamber with time T_2 is added to a recording made in a room with a period T_1 seconds, the resulting sound energy decay is given by the equation

$$10 \log \frac{E}{E_n} = 10 \log (T_1 e^{-14t/T_1} - T_2 e^{-14t/T_2}) - 10 \log (T_1 - T_2)$$

where E_n is the steady state sound energy just before the signal is terminated in the room, and t is the time.

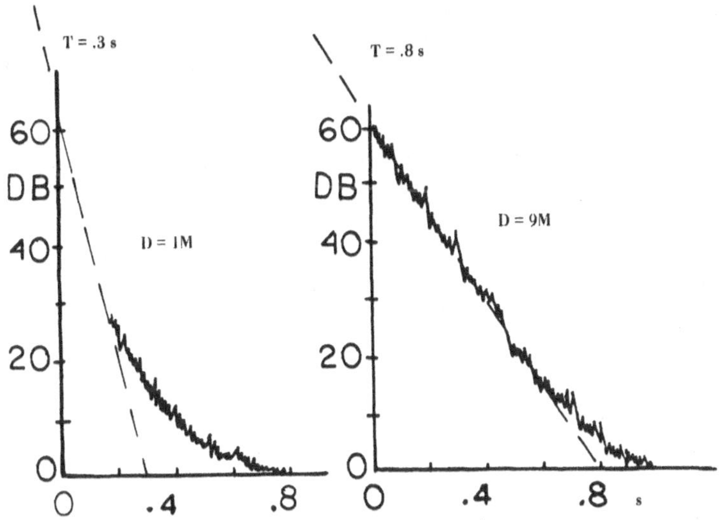

Fig. 4.6 In reverberation time measurements when the microphone is just a short distance (1 m) from the sound source at the time the steady-state signal is suddenly terminated, the initial reverberation time is shorter than the time obtained at a greater distance (10 m) from the loudspeakers. Tests refer to a frequency of 250 Hz.

The resulting sound decay of the two superimposed reverberation periods is not strictly logarithmic in the way T_1 and T_2 by themselves are, but approaches such a curve the greater the difference is between the two periods.

An approximate value for the resulting reverberation time of the two superimposed ones is given by

$$T_3 = (T_1^3 + T_2^3)^{1/3}$$

The space required for reverberation chambers is often far more costly than synthetic reverberators, either of the two-dimensional plate type, like the Swiss EMT units, or else magnetic loop or digital units. However, considering the multiplicity of reflections in a reverberation chamber and the numerous phase relationships between these reflections, it is hard to find a wholly equal substitute for such live enclosures. At any one point in a room, with the loudspeaker located in one of the walls, there are 6 first reflections; 18 second reflections, 38 third reflections, etc. The number of reflections of order n is given by $N = 4n^2 + 2$. Thus there are, for instance, 402 reflections of the tenth order. Such a complicated state of affairs is difficult to duplicate mechanically with a high degree of "authenticity."

4.3 LOW-FREQUENCY SLAT ABSORBERS

To reduce the low-frequency reverberation time in a room without materially altering its high-frequency period, it is desirable to employ specially designed Helmholtz resonators. A popular one is known as the slat absorber, sometimes also called slot absorber, after the use of wooden slats spaced a distance apart from each other, thus forming slots.

Fig. 4.7 shows the principal dimensions of such a low-frequency absorber placed at an angle to a vertical wall to avoid a too sharply tuned absorber. The resonance frequency of such a Helmholtz absorber is, for a first approximation, given by

$$f_o = 54.3\left(\frac{A}{dV}\right)^{1/2}$$

$$= 5480\left(\frac{A'}{d'V'}\right)^{1/2}$$

$$= 2160\left(\frac{A''}{d''V''}\right)^{1/2}$$

where

d = thickness of slat (m)
d' = thickness of slat (cm)
d'' = thickness of slat (in.)
V = volume of space between wall and slats (m^3)
V' = volume of space between wall and slats (cm^3)
V'' = volume of space between wall and slats (in.^{3})
A = area of slot (m^2)
A' = area of slot (cm^2)
A'' = area of slot (in.^{2})

To facilitate the calculations of the slat absorber, the percent open area,

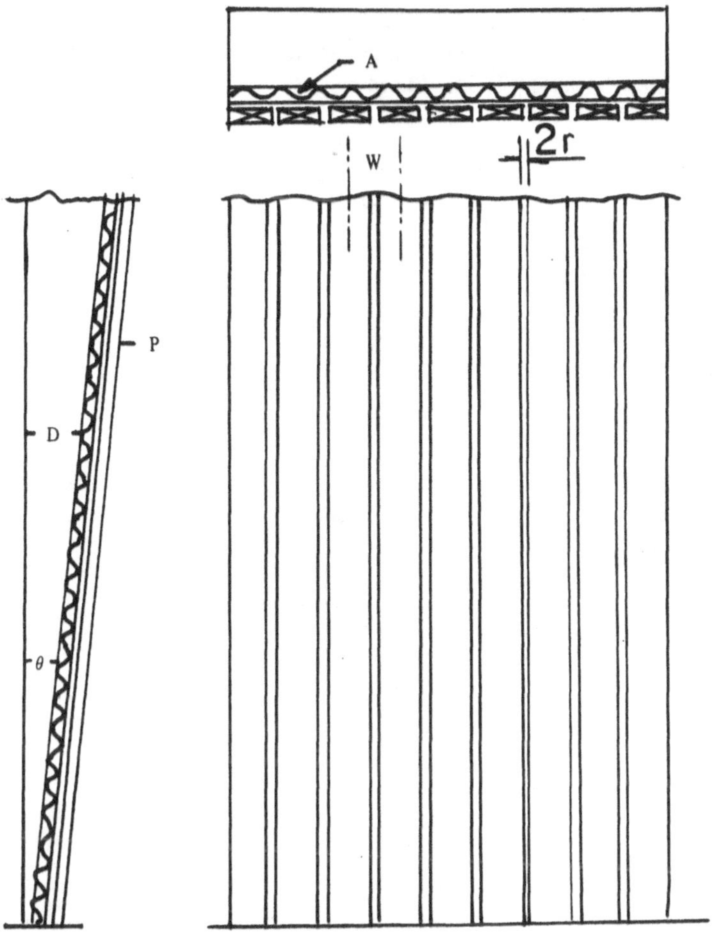

Fig. 4.7 Principal dimensions of slot absorber. A: absorbent material; D: depth of unit; θ: vertical angle of slats

e, is introduced for slots of unit length. Thus (see Fig. 4.8)

$$e = \frac{2r}{w + 2r}$$

where

$2r$ = width of slot (in. or cm)

w = width of slat (in. or cm)

As long as the same type of units are used for the quantities $2r$ and w, e is correct, because it is dimensionless.

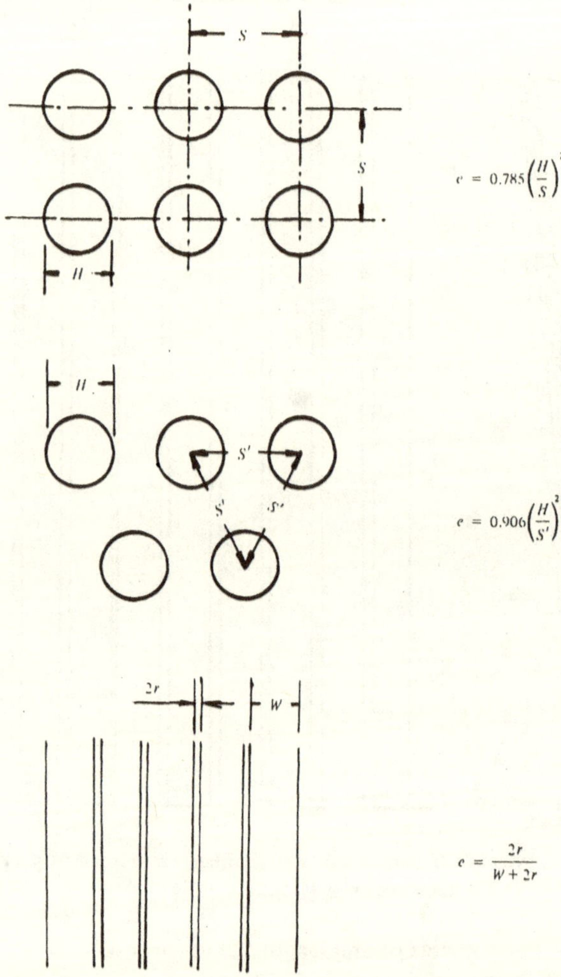

Fig. 4.8 Percent open area of perforated and slotted panels

If an absorber of width W is considered which contains one slot and the two halves of the adjoining slats, the equation

$$\frac{A_o}{V_o} = \frac{e}{D}$$

applies where

D = depth of air space (in. or cm)

A_o = area of slot for unit length of slot (in.^{2} or cm^2)

V_o = volume of air-space behind slats for unit length of slot (in.^{3} or cm^3)

The resonance frequency of the slat absorber is

$$f_o = 2160\left(\frac{e}{dD}\right)^{1/2} = 2160\left[\frac{2r}{dD(w+2r)}\right]^{1/2} \quad r,d,D,w \text{ in in.}$$

$$= 5480\left(\frac{e}{d'D'}\right)^{1/2} = 5480\left[\frac{2r'}{d'D'(w'+2r')}\right]^{1/4} \quad r',d',D',w' \text{ in cm}$$

For a more accurate determination of the resonance frequency, the thickness or depth of the slot (essentially the thickness of the slat) should be replaced by the effective depth d_o. This term, however, is a function of frequency, the cross-sectional shape and area of the slot, and the acoustic properties of the sound-absorbent behind the slots. Hence, the exact effective length must be determined experimentally. For estimative purposes, d_o may be made $1.2d$.

In the references provided at the end of this article, more accurate formulas are given for slot absorbers, but only for such as have no acoustic material behind the slats. In the experience of this writer, the use of $1.2d$ for d_o has resulted in resonance frequencies which agree well with tests when the slot absorber contains 1 in. (2.54 cm) thick fiberglass board behind the slats.

Fig. 4.9 shows the effect of placing the acoustic material at a distance from the slats, instead of in contact with them. This results in a sharper absorption characteristic for the absorber.

Fig. 4.10 illustrates the effect of irregular slots, while maintaining contact between acoustic material and the slats. This results in a broader absorption characteristic.

By slanting the absorber against the wall, a similarly broader absorption characteristic results, compared to that obtained with a parallelepiped air-space between wall and slats.

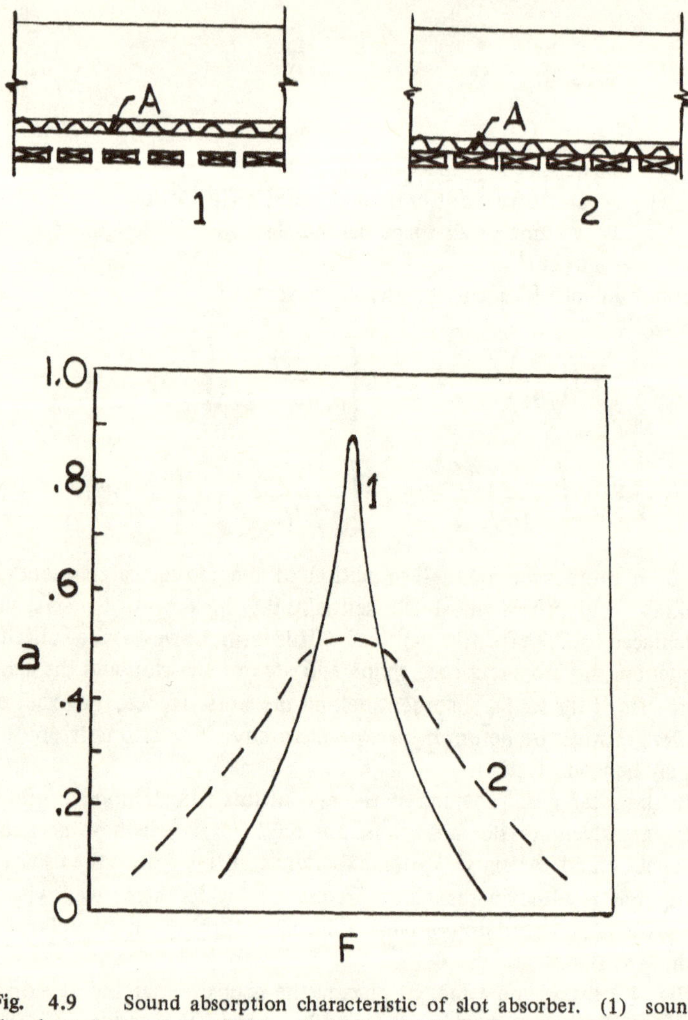

Fig. 4.9 Sound absorption characteristic of slot absorber. (1) sound absorbent A spaced at a distance from slats; (2) A in contact with slats. a: absorptivity.

Below are the requirements for a sharp and a broad characteristic:

Sharp Characteristic	*Broad Characteristic*
Identical slots	Slots of various width
Identical slats	Slats of various width

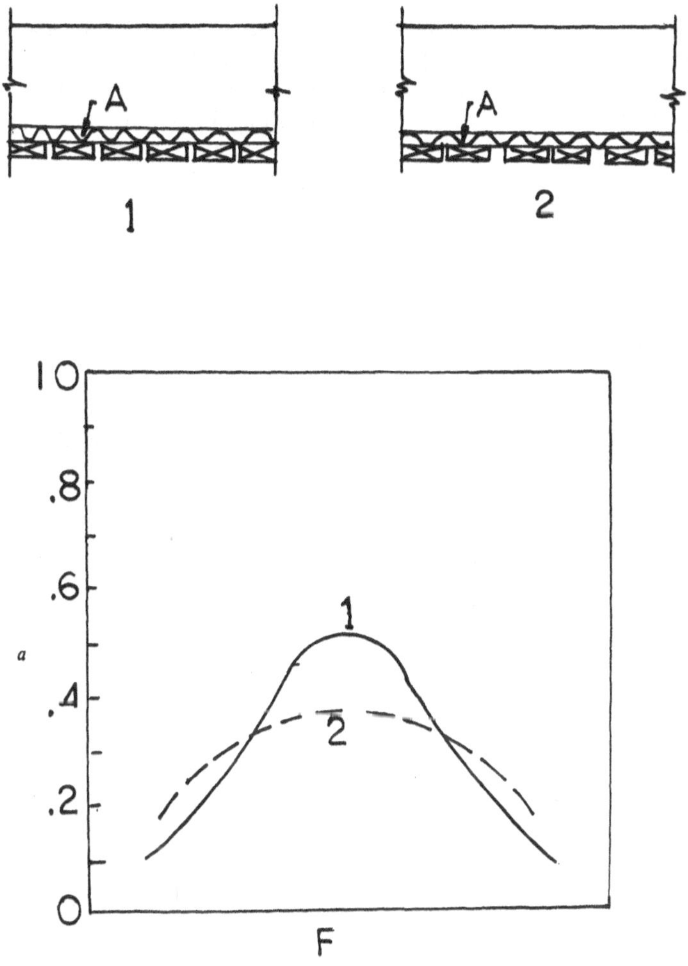

Fig. 4.10 Sound absorption characteristics of slot absorber: (1) slats at constant distance from each other; (2) slats spaced irregularly. A: absorptive material; a: absorptivity.

Space between slats and acoustic material	Acoustic material in contact with slats
Parallelepiped air-space	Prismatic air-space

Variability of slot absorber can also be achieved by nonparallel slots (achievable by the use of nonparallel slats), slats of tapering thickness, and other nonconformities or irregularities. Also a slot absorber can be designed with two sets of slats, one over the other, and so fastened that the exterior slats can slide to cover some or all of the slots of the unit. Changes in the absorption characteristic of a slot absorber can also be achieved by the use of a thick plastic curtain drawn partially over the unit.

Helmholtz resonators are used to advantage when a room has widely spaced low-frequency modes of vibration, brought about either by a small enclosure volume or by an integral or near integral ratio of dimensions. These so-called "eigentones" are of three types: (1) axial, involving reflections from any of the three pairs of parallel walls in a rectangular enclosure; (2) tangential, with reflections occurring parallel along two principal orthogonal directions; and (3) oblique waves, concerning reflections from all six surfaces. Of these modes, the axial ones are frequently the most important. The equations for these vibrations are

$$f_0 = \frac{565n}{L} = 172n/L' \qquad f_0' = \frac{565n}{W} = 172n/W' \qquad f_0'' = \frac{565n}{H} = 172n/H'$$

where L, W, and H are the length, width, and height of the room expressed in feet, n is any integer, and L', W', and H' are the length, width, and height of the room in meters.

Fig. 4.11 shows a studio 15.5 ft (4.7 m) high, 22.1 ft (6.74 m) wide, and 31.4 ft (9.57 m) long. It has the following axial modes along the:

Length	*Width*	*Height*
18.0 Hz		
	25.6 Hz	
36.1 Hz		36.5 Hz
	51.3 Hz	
54.1 Hz		
72.1 Hz		72.7 Hz
	76.7 Hz	
90.2 Hz	102.2 Hz	
108.2 Hz		108.8 Hz
126.3 Hz		
	127.8 Hz	

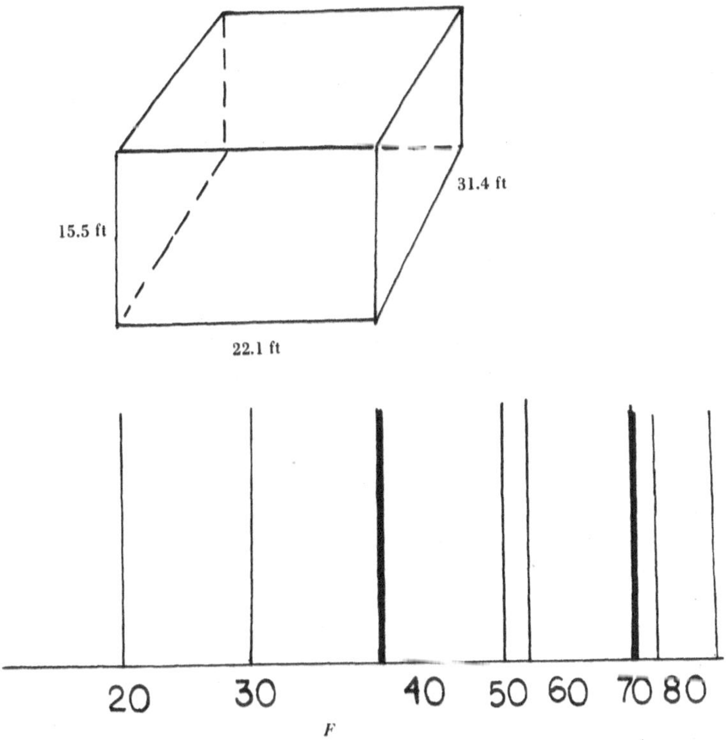

Fig. 4.11 Number and frequency distribution of axial modes of vibration for room shown on top. F: frequency in hertz.

Fig. 4.11 also shows the distribution of the axial modes along a logarithmic frequency scale. It is seen that the widest spacing is between the frequency of 54.1 and 72.1 Hz (18 Hz) and that there are double modes near 36 and 72 Hz. This condition can cause coloration (spectrum- or pitch-change), particularly during reverberation.

Assume that a Helmholtz absorber is installed in the form of a slat unit employing 2 in. × 4 in. (5 cm × 10 cm) slats with an air-space $D = 10$ in. (25 cm) behind the slats, and that the unit should be most absorptive at 72 Hz. What should the width of the slots be?

$$f_0 = 2160 \left[\frac{2r}{d_o D(w + 2r)} \right]^{1/2}$$

$$72 = 2160 \left[\frac{2r}{2.4 \times 10 \times (4 + 2r)} \right]^{1/2}$$

By solving the above equation for $2r$ we obtain 0.1 in. (0.25 cm) for the width of the slots.

In the design of a slot absorber precautions should be taken to avoid a depth dimension D which is equal to a half wavelength of the desired frequency of resonance. The reason for this is that the wave reflected from the hard wall in back of the slats would interfere with the high particle velocity at the interior surface of the slot, since it would be 180 degrees out of phase. Similarly, when D is equal to a quarter wavelength of the resonance frequency, the air stream at the "mouth" of the resonator would be additionally higher because the particle velocity of the reflected sinusoidal wave is highest at a quarter wavelength from the hard wall.

When the airspace behind the slats is subdivided by partitions, so that a number of resonators come to exist with smaller volumes, the resonance frequency of each resonator so created is higher than that of the total unit (with equal and uniform slots). This is evident from the fundamental equation of a Helmholtz resonator whose frequency is inversely proportional to the square root of the volume behind the slats.

The use of vases in a studio for sound-absorptive purposes must be carefully considered, because such resonators may actually prolong the reverberation in a room. It was this type of vessel, generally made of bronze, which the Greeks used in their open-air amphitheaters so that "the sound is strengthened and beautified" (Vitruvius). Such prolonged reradiation of the signal from metal jars resembles that of a bell, which when struck may continue to vibrate for a period of time ranging from a few milliseconds to possibly a second or more.

The slats for a slot-absorber do not have to be a parallelepiped, that is, have a rectangular cross section. Fig. 4.12 shows various cross-sectional shapes for the slats. The slats shown on the bottom of the figure are similar to those used in the Herrenhaus-Sitzungssaal of the Parliament in Vienna, Austria. The rectangular shape of the slats is not the best, particularly when a large wall area is covered with them, since such a surface can give rise to high-frequency echoes in the manner of perforated hardboard or transite. Parenthetically, the common perforated 1/8 in. (0.32 cm) thick Masonite with a 10 percent open area, and backed by 1 in. (2.5 cm) absorbent material is not a low-frequency but more nearly a mid-frequency and even high-frequency absorber, depending on the spacing between acoustic material and Masonite, the size holes, and type of absorbent.

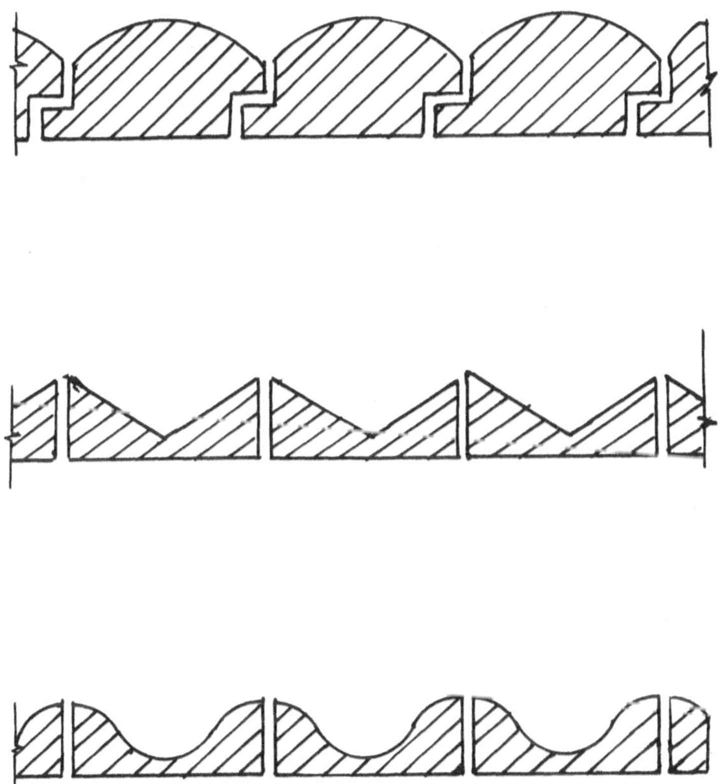

Fig. 4.12 Decorative slats for low-frequency Helmholtz-type sound absorber

For practical design purposes, the resonance frequency of a slot absorber may be written as

$$f_0 = 5480 \left(\frac{e}{1.2dD} \right)^{1/2}$$

$$= 5000 \left(\frac{e}{dD} \right)^{1/2}$$

where

d = slot depth (cm)
D = depth of space between slats and hard wall (cm)
e = percent open area
$= \dfrac{2r}{w + 2r}$
$2r$ = width of slot (cm)
w = width of slat (cm)
V = unit air volume behind slats
$= D(w + 2r) \times 1 \ (\text{cm}^3)$

From a knowledge of the desired resonance frequency, the above equation can be solved for any of its factors. Thus

$$e = 4 \times 10^{-8} dDf_o^2$$

$$D = \frac{2.5 \times 10^7 e}{df_o^2}$$

$$d = \frac{2.5 \times 10^7 e}{Df_o^2}$$

$$2r = \frac{4wdDf_o^2 \times 10^{-8}}{1 - 4f_o^2 dD \times 10^{-8}}$$

$$V = WD \times 1$$

$$= (w + 2r)D$$

$$= \frac{2.5 \times 10^7}{f_o^2 \dfrac{d}{2r}}$$

A more accurate expression is

$$V = \frac{3 \times 10^7}{f_o^2 \left(\dfrac{d}{2r} + 6.62 - 1.462 \log 2rf_o\right)}$$

In the late 1970's there appeared in the recording studios a type of absorber called a "trap." It is essentially a deep airspace filled with panels of fiberglass board spaced an inch or two apart, thus forming deep vertical slots. This type of absorber is like an "acoustic sink," absorbing all frequencies to a very large extent. Such a trap may be placed near a drum or trumpet, thus absorbing a great deal of the signal coming from the instrument, thus avoiding high-level signal transmission into adjoining microphones in multiple track recording. Hence such terms as "drum trap" or "trumpet trap" exist where these instruments are placed in a studio. Such units may be made into deep pits in the floor, covered with an iron grill work to prevent accidents.

4.4 BASS TRAPS

In the late 1970's some recording companies installed the so-called bass traps in the floor, walls, or ceiling of their studio. These are pits or recesses several feet deep, lined with 1- or 2 in.-thick vertically-suspended, fiberglass panels and covered first with a horizontal layer of the material and then with either a grid when the trap is in the floor or with a thin cloth when it is in a wall or on the ceiling.

When the recess depth represents a quarter wavelength of the incident bass note (and odd multiples thereof) maximum sound absorption occurs at that frequency. The reason for this is that the air particle velocity at a quarter wavelength from a hard reflecting surface is a maximum. At that position the molecules move in and out of the horizontal layer so quickly that a considerable amount of the acoustic energy of the wave is converted into heat within the frictional interstices of the material. Other notes are also absorbed as their waves pass between the spaced fiberglass panels, strike the soft bottom layer, and return again through a highly absorbent environment.

The table below gives the quarter wavelength of various bass notes, and thus the required trap depth for this frequency of "suppression:"

	Depth	
f	m	ft
30	2.86	11.38
40	2.14	7.04
50	1.72	5.64
60	1.43	4.69
70	1.23	4.03
80	1.07	3.52
90	.96	3.13
100	.86	2.82
120	.72	2.35

When a band of bass notes is absorbed, the bottom of the pit should be slanted. As an example, when the octave between 40 and 80 Hz is to be expunged from the incident signal, one end of the hole in the ground should have a depth of 7 ft and the other end should have one of 3.5 ft. For such wide-band absorption the cross-sectional area of the trap should be larger than when only one sinusoid is to be effectively absorbed.

The very high absorption capability of such a fiberglass-lined floor, wall, or ceiling chamber, as far as the whole room is concerned, is due to the fact that precisely where the air-particle velocity of the reflected wave is a maximum, the sound pressure is a minimum, being 90 degrees out of phase with the particle velocity. This is because at the bottom of the pit the air particles can not move at all because they are facing an immobile barrier and because of the increased pressure, a doubling of the amplitude results. For this reason the face of the trap represents a vacuum, drawing "sound rays" into the cave which otherwise would have traveled elsewhere, as diagrammatically illustrated on Fig. 4.13.

The purpose of a bass trap is twofold–to absorb floor reflections for the bass notes of the violas, drums, and other instruments with predominantly low-frequency output when the instrument is over the trap, and to lower the music level at adjoining instruments. The first purpose is to pre-

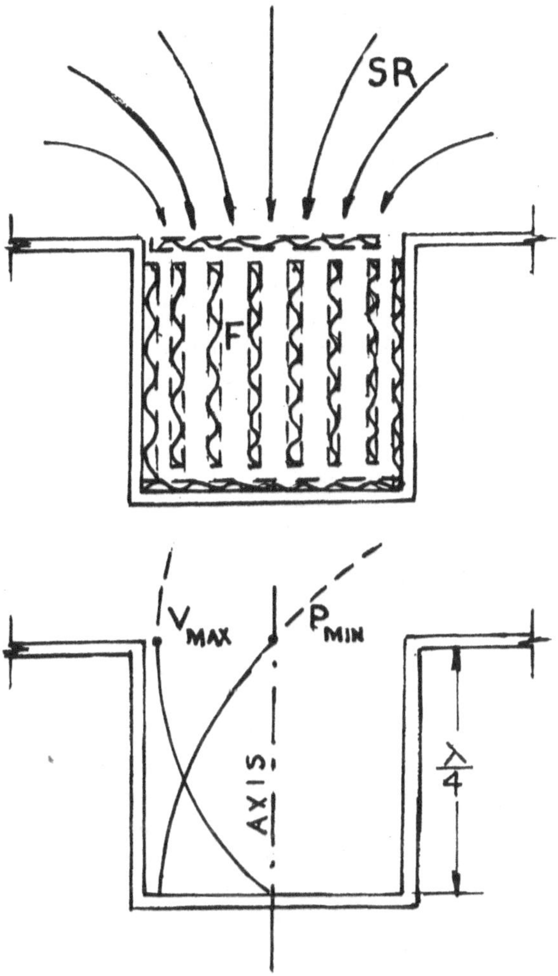

Fig. 4.13 Sound trap. SR: sound rays; F: fiberglass board; V: velocity variation of air particle; P: pressure wave

vent such notes from entering adjacent microphones in multiple track recording, and the second purpose is to provide greater acoustic comfort for the adjoining player. Drum notes, for instance, can reach a sound pressure level of 100 dB at a distance of 1 m or 3 ft, and violas, too, can generate acoustically uncomfortable levels, particularly for nearby violinists. An incidental purpose of a trap is to absorb real low notes generated anywhere in the studio.

For these reasons drummers are often placed in drum cages (see Fig. 4.14) and pianists in a similar enclosure, away from the band, often to the discomfort of these performers.

By the experience of some recording engineers, the use of "flats" between instrument sections has often led to insufficiently clean tracks in multiple track recording. Often this is due to wheeled partitions which are either not tall enough, allowing signals to creep over and under these units, or not deep enough so that sound can pass around the front and rear vertical corners of them. Such devices, cheaply or inexpertly constructed, also do not allow a performer to see others in adjacent sections for better timing because the space dividers have no windows. Some engineers consider flats too cumbersome to adjust, and when not in use take up valuable floor space unless storing facilities are planned for them during the design stage of the studio. Finally, as handcrafted items, such units are expensive, and sometimes obscure the view of the mixer intent on watching the performers.

In this connection, the writer would like to cite an experience he had in 1943 when the Columbia Picture Corporation in Hollywood decided to make "The Jolson Story" with Larry Parks. Al Jolson had been commissioned to sing the important songs in "The Jazz Singer." He insisted on rendering his vocal numbers within the orchestra, because as a stage performer he had always been surrounded by musicians. So a glass cage was built with a heavily absorbent floor and ceiling, and earphones were offered to him for the purpose of singing in synchronism with the instrumental music. He would not wear them and surprisingly, the little sound which filtered through the glass was enough for him to stay in time, and it could not be heard on the track.

There are two disadvantages connected with the use of bass traps. One is that the music sounds less lively, rich, and full, and the second is that the performer over the trap feels uncomfortable, as if he or she were suspended in midair without any surrounding reinforcing sounds.

To analyze the effects, the physics of sound must be examined. In 1931 G.W. Steward, Professor of Physics at the State University of Iowa, published the "Acoustic Uncertainty Principle."* It is similar to Heisenberg's uncertainty principle in quantum mechanics which states that an electron

* Steward, G.W. "Problems suggested by an Uncertainty Principle in Acoustics," Jl. Acous. Soc. America V.2, No. 3, Jan. 1931, page 325.

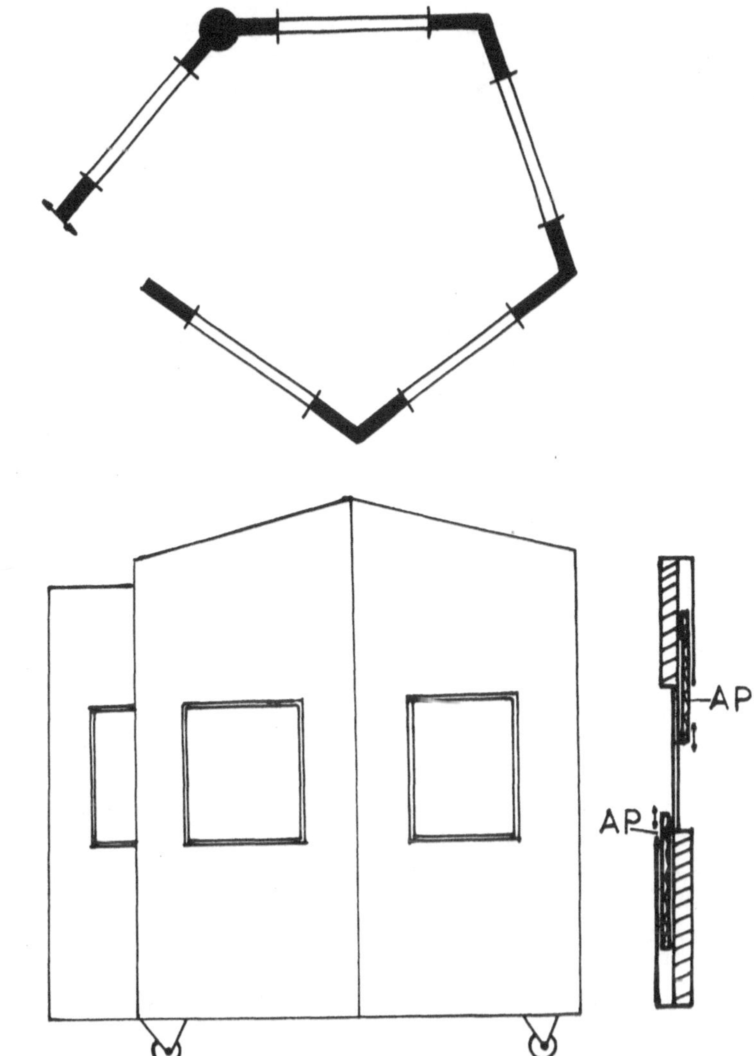

Fig. 4.14 Drum cage on wheels. AP: moveable acoustic panels over windows.

can never be seen under a microscope, because the impact of the light ray upon the subatomic particle moves it so much that the reflected light ray in the eyepiece can show up as only a blur. By the acoustic uncertainty principle a similar inability to recognize the components of a complex sound wave exists. Stated mathematically, both the Heisenberg and the Steward uncertainty principle may be expressed as follows:

$$\Delta f \times \Delta t = 1$$

where f refers to the frequency in the acoustic case and to the total energy of the moving particle in the case of quantum mechanics, t represents the time in both cases, and Δ is the class interval.

As an example of the acoustic principle of uncertainty, consider a musical note. One does not hear a single frequency tone but a pulse which by Fourier analysis is found to consist of a number of sinusoids. There is a frequency f_1 of maximum intensity and a range of higher and lower frequencies with different intensities. When one tries, aurally, to identify f_1 there is difficulty in recognizing it because of the presence of the other components. As an example, assume that a determination of a 1000 Hz f_1 with 1 percent accuracy is required. This means that f_1 must be evaluated within 10 Hz. By the acoustic principle of uncertainty, the required time interval is 0.1 s because $10 \times 0.1 = 1$. When the frequency interval is to be only 3.3 Hz, the required time interval becomes 0.3 s, because $3.3 \times 0.3 = 1$. In other words, the accuracy of the frequency definition increases with time. But there is no way in an anechoic environment to increase this time, because there are no "fusing" reflections. However, by placing a reflective surface near an instrument, this time interval is increased by the minutely delayed reflection, and the reflection provides, advantageously, a greater tone intensity at the ear of the performer. Musicians are very conscious of this effect and call it "attack" or "tonal force." This is also one of the reasons why we like to sing in a shower, although in this case we also have the effect of a longer reverberation time.

Nonpercussion instruments are difficult to recognize by their steady-state timber. Identification requires the perception of their starting transients, or attack, of the notes. The note growth is generally described as the level increase to within 3 dB of maximum level. In duration it varies from 10 to 100 ms, depending on the instrument, the player, and the note played. This attack contains a number of important characteristic blips, or short sections of amplitude modulation, which like the tone pitch require a time period for recognition. Fig. 4.15 gives a graphic illustration of an attack transient.

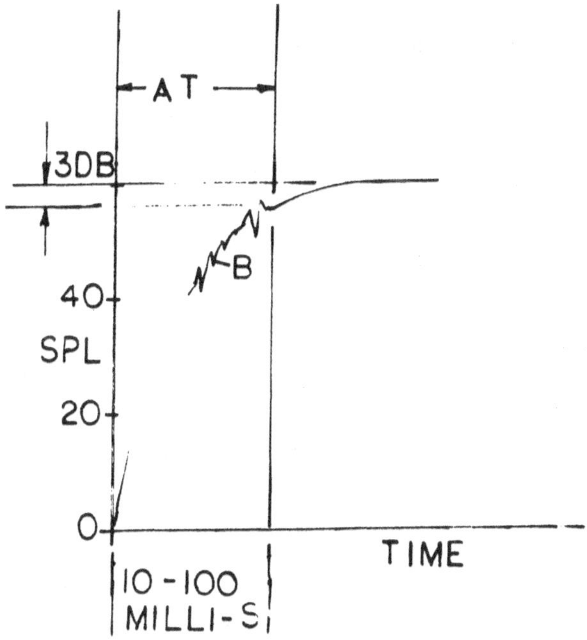

Fig. 4.15 Note growth of a nonpercussion instrument. AT: attack transient; B: blips or short sections of amplitude modulations.

Now what about acoustic comb filter effects when there is a reflecting surface near an instrument? What about the sounding board of a piano which reflects the action of the strings? Or better yet, what about the twenty violins in a symphony orchestra? At the microphone their signals are all out of phase with each other, and yet their rendition is better than that of a single violin because of these phase relationships. At the beginning of sound-on-film recording, a producer once thought he could save money by having the recordist turn up the amplifier level some 13 dB in imitation of the signal strength of 20 violins. The effect was very discouraging when the photographed scene showed a violin ensemble where the distant performers were not musicians but mimicking extras. So let us not ignore the facts which have made music rich in appeal, even when acoustic spectrometry appears to show an anomaly. A great deal must still be learned about the ear and, especially, about the human brain which analyzes what is heard.

A theorist might object to such bass traps for reasons other than those already mentioned. He might say that a studio with such a trap is no longer a room, a three-dimensional enclosure, because the sound-absorbent pit represents a discontinuity of the bounded space, where sound can enter but not return. By the G. Millington equation, this theorist might say, it is not even possible to calculate the reverberation time of the studio. This is for the reason that the equation contains the term $S \log (1 - a)$, where S is the cross-sectional area of the trap and a is its unity absorption coefficient, which would become minus infinity. Actually, by our method of absorption measurements, a would turn out to be larger than unity, for the reason that some of the normally diffused signal would be "sucked" into the hole which otherwise, would travel in straight lines about the enclosure. Lack of uniform distribution of the sound energy would result in a nonexponential decay of the signal on sudden stoppage. These are fine points, which in studios with a reverberation period of 0.2 to 0.4 s would probably not be noticed by the majority of performers. They are presented here for the sake of an analysis of the subject of bass traps.

Another type of low-frequency sound absorber consists of a nonporous panel, like a sheet of plywood or plasterboard, spaced at a distance from a hard wall, with the interspace filled loosely with an absorbent like fiberglass or felt. The panel is considered a mass connected to a spring (the elasticity of the trapped air), and has a resonance frequency of

$$f_0 = \frac{170}{(Md)^{1/2}}$$

$$= \frac{60}{(M'd')^{1/2}}$$

$$= \frac{190}{(M''d'')^{1/2}}$$

where

M' = surface density of panel in kg/m^2
M = surface density of panel in lb/ft^2
M'' = surface density of panel in g/cm^2
d' = distance between panel and hard wall (m)
d = distance between panel and hard wall (in.)
d'' = distance between panel and hard wall (cm)

In the case of a stretched circular membrane, the fundamental resonance frequency is

$$f_0 = \frac{0.0766}{D'} (T'/M')^{1/2}$$

$$= \frac{0.766}{D} (T/M)^{1/2}$$

where

M' = surface density of membrane (kg/m^2)
M = surface density of membrane (g/cm^2)
T' = tension of membrane (N/M)
T = tension of membrane (dynes/cm)
D' = diameter (m)
D = diameter (cm)

In the case of a square membrane, like an oil painting in a frame,

$$f_0 = \frac{0.792}{a} (T/M)^{1/2}$$

where a is the length of a side of the membrane, and T and M are the same as for a circular membrane.

Such membranes are often very good low-frequency absorbers, as are tents and even paper pie plates cemented to a wall. At one time there was on the market an acoustic tile in the form of an inverted square paper pie plate which could be cemented to the ceiling to achieve effective noise reduction in rooms where a low-frequency hum or rumble existed. These tiles were known as the "Heerwagen" diaphragm.

The walls of a room can often be likened to sheets thinner than plates. A similar complicated situation exists for boxes and containers of various types. Violins made with thin spruce wood for the bellies exhibit more low-frequency response than violins with thicker bellies. Violins made of aluminum with a surface density equal to thick spruce panels also lack low-frequency response. In general, the motions of such sheets or parts of them under the influence of a complex note show motions which consist of the superpositions of many modes of vibration. For this reason also, violins can only be rated qualitatively, by ear, and not by spectrum analysis. The application of even a thin coat of varnish to a violin may result in a noticeably better tonal response. However if the varnish is slightly thinner or thicker than optimal, the tonal response of the violin can be worse.

It is seen, therefore, that bass traps are not helpful to the musician, who must feel suspended in midair when placed over a bass trap, particularly when the wall behind him as well as the ceiling above him have been rendered anechoic. Nor are bass traps conducive to many types of musical renditions because the absence of nearby reflective surfaces, by the acoustic uncertainty principle, reduces tone definition as well as the type of immediate tonal response so much sought after by the amateur vocalist. These traps, however, help the recording engineer in acquiring clean tracks, when that is the most important factor in album recording.

Chapter 5

Air-Conditioning System Noise Limits

5.1 GENERAL

Fig. 5.1 shows the family of *PNC* (Preferred Noise Criterion) curves currently employed in architectural acoustics for describing permissible noise level limits in various rooms.

The recommended *PNC*-20 curve should apply to all significant recording and listening positions in the enclosure. In large spaces there may be a considerable difference in the A-weighted sound level near the diffuser at the ceiling and the allowable limit on the floor. When microphones are intended to be stationed near the ceiling, the noise level there should also meet the specified *PNC* rating. This means that the noise level at the floor may be lower than at the ceiling, because of the inevitable noise level reduction with distance. On the other hand, when the A-weighted sound level 3 ft (0.91 m) from a diffuser is X dB-A, and there are N such noise sources at the ceiling of height H, the total noise level near the floor will be $X + 10 \log N - 10 \log (H/3)$, when the room is "dead." Thus, when there are 10 diffusers on the ceiling and the room height H is 30 ft (9.1 m) the increase in the noise level due to the 10 ceiling sources just about equals the decrease of the noise levels with distance.

The mechanical engineer or the air-conditioning system contractor for a studio is usually furnished the sound power levels of the fans or compressors at the intake of the duct system. Knowing the duct lengths from the equipment to the diffusers, he can calculate the required noise reduction in the ducts to achieve the desired noise levels at the diffusers. These calculations have been amply described in the literature available from ASHRAE, the American Society of Heating, Refrigeration, and Air-Conditioning Engineers, and are not discussed here.

The ARI (Air-Conditioning and Refrigeration Institute), in an attempt to simplify noise level calculations for air-conditioning system equipment, has established a rating system described in its Standard 275-69 of 1971.

Recommended Background Noise Criteria for Various Rooms

Type of Enclosure	PNC	A - Weighted Sound Level
Recording Studios, Concert Halls	15 - 20	25 - 30
Theatres	20 - 25	30 - 35
Homes	25 - 30	35 - 40
Offices	30 - 35	40 - 45
Stores	35 - 40	45 - 50

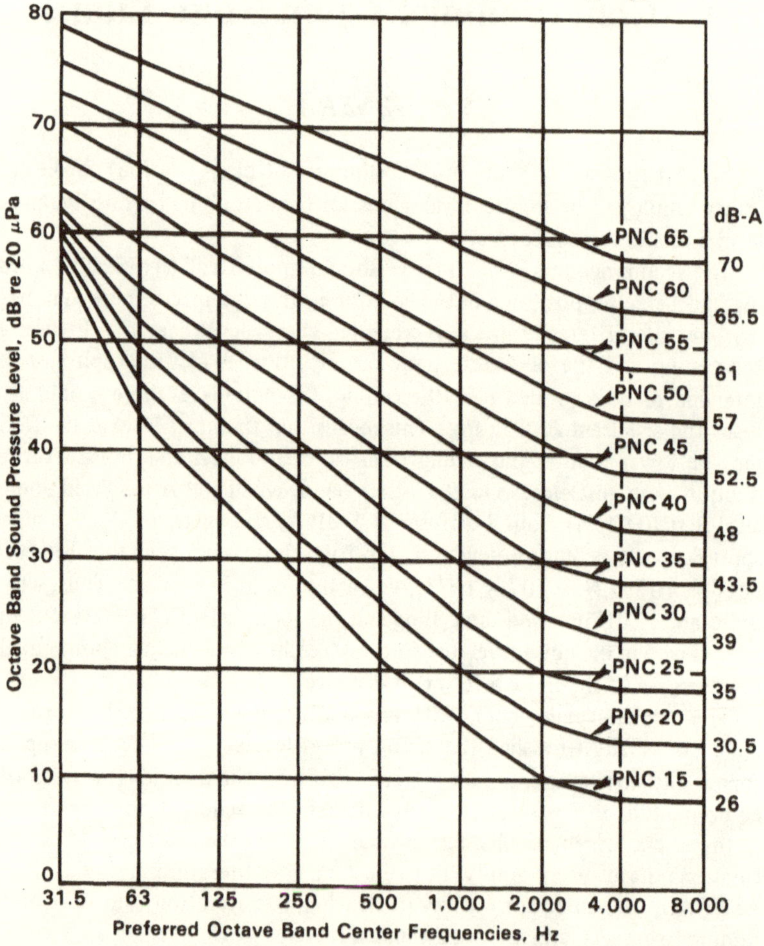

Fig. 5.1 Preferred Noise Criteria Curves. Each curve is a code for specifying the noise level characteristic, or spectrum, of noise in rooms.

By means of a nomograph one can calculate, from a given ARI number, the A-weighted sound level at a desired distance from the machinery. The information can also be obtained by the following equation (see Fig. 5.2)

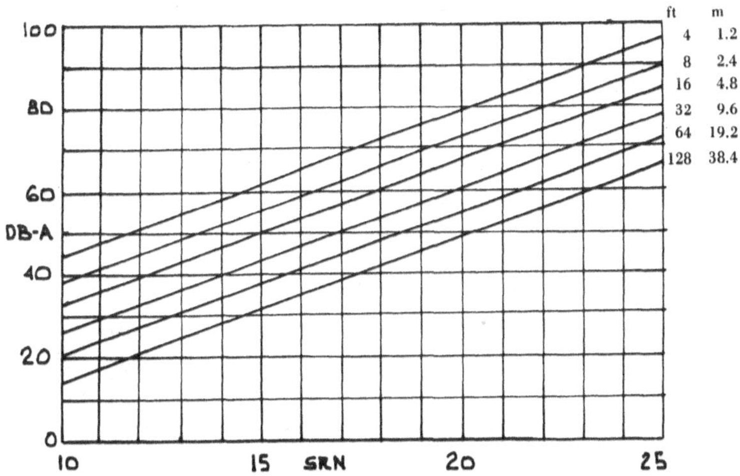

Fig. 5.2 Variation of A-weighted sound level with *SRN*, the noise equipment rating number, for various distances between machinery and observer.

$$SL_A = 3.4\,SRN - 20\log D + 17.4 = 3.4\,SRN - 20\log D' + 7.1$$

where

SRN = sound rating number
D = distance (ft)
D' = distance (m)

As an example, consider an air-conditioning system machinery with an ARI rating of $SRN = 20$. What is the A-weighted sound level 50 ft from the sound source?

$$SL_A = 3.4 \times 20 - 20\log 50 + 17.4$$
$$= 51.4 \text{ dB-A}$$

It should be realized that the scheme proposed by ARI represents an approximation, since it does not consider the directional pattern of the sound source. When the sound power level of an air-conditioning system equipment is known, a more accurate equation for the A-weighted sound level at a distance is

$$SL_A = WL - 20 \log D + 10 \log Q - 0.7$$
$$= WL - 20 \log D' + 10 \log Q - 11$$

where

WL = A-weighted sound power level referred to 10^{-12} watts
Q = directivity factor
D = distance in feet
D' = distance in meters

But even this equation is somewhat of an approximation, since the sound level in the open may decay less than 6 dB for double the distance, and may diminish only 4.5 dB, in which case the factor for the log term of the distance is not 20 but only 15.

Intimately connected with the maximum noise level limits in a room are the refrigeration capacities of its air-conditioning system. There are well-known formulas for such requirements when the building is of the hotel, office, or apartment house type. There is a scarcity of such information when the structure is a TV or motion picture studio stage.

The engineer called upon to specify the air-conditioning system load for entertainment program enclosures should base his calculations on the preferred operating conditions of the production and engineering departments of the studio under consideration. What he specifically needs to know is: the maximum *NPA*, or net production area, in square feet or square meters, which is to be illuminated with the light intensity of incandescent lamps in footcandles or luxes necessary for good picture definition and high picture signal-to-noise ratio; the tungsten illumination (generally 3200 K), the prevailing focal lengths of the camera lenses in use at the studio, the type of pick-up tube (image orthicon, vidicon, plumicon, etc.); and the desired *f*-stops of the lenses to achieve the necessary depth of field for the picture.

Fig. 5.3 shows the required cooling capacity in tons and the associated electric power as a function of the net production area and the specified illumination in footcandles.

These refrigeration capacities must be modified when other significant conditions exist, such as an audience in the room, animals, projection equipment for front- or rear-screen process work, and others.

Also, the lighting load depicted in Fig. 5.3 is that only for the lighting within the confines of the cyclorama screen, and does not include the

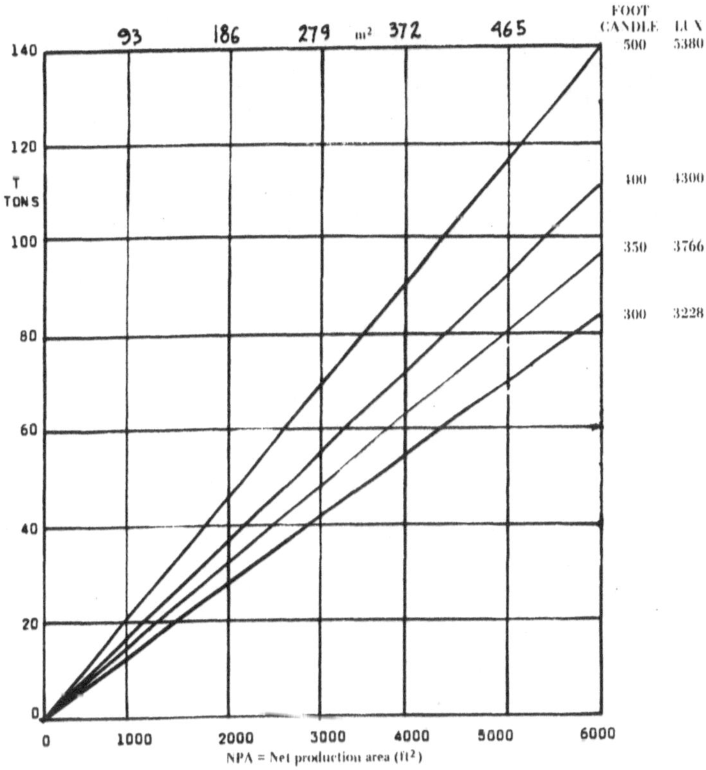

Fig. 5.3 Variation of refrigeration capacity in tons of a studio air-conditioning system with the net production area for various illumination requirements. Multiply tons of refrigeration by 3.05 to obtain megacalories per hour.

screen's own lighting load. For a 20-ft (6.1 m) high cyclorama screen which is to be lit both from the top and bottom, 2 kilowatts of power should be allowed per lineal foot of screen (6.56 kilowatts per lineal meter) when incandescent lamps are used.

The equation for the total refrigeration capacity of the studio's air-conditioning system for lighting is given by

$$T = 4.66 \times 10^{-5} \times FC \times NPA + 0.28L \qquad \text{(tons of refrigeration)}$$

$$= 0.466 \times 10^{-5} \times LU \times N'P'A' + 0.92\,L' \qquad \text{(tons of refrigeration)}$$

$$T' = 14.2 \times 10^{-5} \times FC \times NPA + 0.28L \qquad \text{(megacalories per hour)}$$
$$= 1.4 \times 10^{-5} \times LU \times N'P'A' + 0.92L' \qquad \text{(megacalories per hour)}$$

where

FC = required illumination in footcandles
NPA = net production area (ft^2)
L = length of cyclorama screen (ft)
LU = required illumination in luxes
$N'P'A'$ = net production area (m^2)
L' = length of cyclorama screen (m)

Thus, when 400 footcandles of illumination are required within a 3000 ft^2 net production area surrounded by a 200-ft long cyclorama screen of 20-ft height, the refrigeration capacity is

$$T = 4.66 \times 10^{-5} \times 400 \times 3000 + 0.28 \times 200$$
$$= 112 \text{ tons} = 338 \text{ megacalories per hour}$$

The service power is

$$W = 7.15T$$
$$= 7.15 \times 112$$
$$= 800 \text{ kW}$$

For a 120-V ac service, the current is $8 \times 10^5/120 = 6666$ A.

The previous figure of 2000 W of electric power for the lighting load per lineal foot of a cyclorama screen 20 ft (6.1 m) high is sometimes held to be minimal. At a SMPTE conference in Los Angeles on April 22, 1974, a figure of 3000 W was held to be preferred. At the same conference, Dr. William E. Glenn* recommended a fluorescent lighting system with significant power savings, a more uniform light distribution, and greater color saturation.** Whether fluorescent lighting of the cyclorama screen is acceptable to the studio depends, of course, on the experience and preferences of its technical staff.

When the cyclorama screen is higher than 20 ft (6.1 m), the above figure for the incandescent lighting power load is inadequate, and the load should

* CBS Laboratories, High Ridge Drive, Stamford, CT, 06905

** See "A New Method for Cyclorama Lighting" by William E. Glenn and Salvatore J. Bonsignore, September 1974 *Journal of the SMPTE*, V. 83, No. 9, p. 722.

be figured more nearly on an area basis. A good figure to use, in the experience of this investigator, is 250 W/ft^2 (2500 W/m^2) of cyclorama screen.

The electric engineer for the planned project should also consider power requirements for projection rooms, video and audio equipment, and even wind machines. An important consideration involves the conversion of alternating current to direct current for various processes involved in production.

In the production of motion picture and TV programs it is often necessary to generate fog, steam, smoke, etc. on the set. These fumes must later be removed from the room and cannot be run into the air-conditioning system where they would pollute the air in other enclosures fed by the same system. Hence a ventilating system is required on the stage, which exhausts the smoke-filled air into the atmosphere. To protect the stage against aircraft flyover disturbances and other noises, the exhaust system should be baffled as shown in Fig. 5.4.

Fig. 5.4 Sound-retardant air-exhaust openings for ventilating system in stage to remove fumes

When rehearsals are to take place during smoke exhaust periods, and relative quietude is necessary while the fan is in operation, it is necessary to also provide a baffled duct below the roof so that both fan noise and aircraft flyover disturbances are absorbed before they can enter the room.

In general it is advisable to have the cool air of the system enter at the upper portions of the room and be returned along the lower portions of the walls. When the air enters at the lower portions of the walls it may be able to bring sets into vibration, which of course is an undesirable condition from a photographic point of view. Also, it has happened that when sparsely clad dancers have to stand around near the walls during rehearsals, they developed goose pimples, with equally undesirable photographic results.

In calculating the sound pressure levels produced by fans at various distances it is best to rely on the sound power level characteristic quoted by the manufacturer of the fan. In the absence of such information, the sound power levels at different frequencies for a fan with a given air volume flow rate, number of blades, and other data, Fig. 5.5 may be consulted. The example quoted at the bottom of the figure is self-explanatory.

Quoted fan power levels generally refer to the total fan power of the unit. At both the inlet and the outlet of the fan, the sound power is then 3 dB less, because of the fan sound power split.

The fan sound power outside the housing is that of the total fan power minus the sound transmission loss of the housing material. For various thicknesses of sheet metal of the housing the following *TL*'s may be deducted:

Sound Transmission Loss dB

Steel Thickness			*Frequency* 63	125	250	500	1000	2000
0.159 cm	0.0625 in.	1/16 in.	11	17	23	29	35	41
0.120 cm	0.047 in.	3/64 in.	9	15	21	27	33	39
0.078 cm	0.031 in.	1/32 in.	7	13	19	25	31	37

The sound pressure levels at the place where the fan is connected to the duct are determined by the duct openings. The equations are

$$SPL = WL - \log S + 10.6$$
$$= WL - 10 \log S'$$

where

S = duct opening (ft^2)
S' = duct opening (m^2)

When $WL = 90$ dBp, $f = 250$ Hz, and $S = 10$ ft^2, then

$$SPL = 90 - \log 10 + 10.6$$
$$= 90.6 \text{ dB}$$

PROCEDURE FOR ESTIMATING THE SOUND POWER LEVELS OF VARIOUS FANS WITH N BLADES

$WL = K + 10 \log Q + 20 \log P + BFI$ dBp

Q = air volume flow rate in ft^3/min
P = static pressure in inches of water
BFI = blade frequency increment in dB
K = parameter, given in table below
BF = blade passage frequency = $N \times$ rpm/60

Fan Type	K 63	125	250	500	1K	2K	4K	8K	BFI
Centrifugal									
Airfoil blade	35	35	34	32	31	26	18	10	3
Backward curved blade	35	35	34	32	31	26	18	10	3
Forward curved blade	40	38	38	34	28	24	21	15	2
Radial blade	48	45	43	43	38	33	30	29	5–8
Tubular	46	43	43	38	37	32	28	25	4–6
Vane axial	42	39	41	42	40	37	35	25	6–8
Tube axial	44	42	46	44	42	40	37	30	6–8
Propeller	51	48	49	47	45	45	43	31	5–7

$WL = K' + 10 \log Q' + 20 \log P' + BFI$ dBp

Q' = air volume flow rate in m^3/s
P' = static pressure in centimeters of water
BFI = blade frequency increment in dB
K' = parameter, given in table below
BF = blade passage frequency = $N \times$ rpm/60

Fan Type	K 63	125	250	500	1K	2K	4K	8K	BFI
Centrifugal									
Airfoil blade	60	60	59	57	56	51	43	35	3
Backward curved blade	60	60	59	57	56	51	43	35	3
Forward curved blade	65	63	63	59	53	49	46	40	2
Radial blade	73	70	68	68	63	58	55	54	5–8
Tubular	71	68	68	63	62	57	53	50	4–6
Vane axial	67	64	66	67	65	62	60	50	6–8
Tube axial	69	67	71	69	67	65	62	55	6–8
Propeller	76	73	74	72	70	70	68	56	5–7

Example: Calculate the estimated sound power level at the blade frequency of a 12-blade fan with a centrifugal radial impeller turning 625 rpm and delivering 50,000 ft^3 of air/min with a 2 in. static pressure drop.

$BF = 625 \times 12/60 = 125$ Hz $\quad K$ at 125 Hz for subject impeller is 45 dB by table. $BFI = 6$ dB

$$\begin{aligned} WL &= K + 10 \log Q + 20 \log P + BFI \\ &= 45 + 10 \log 50{,}000 + 20 \log 2 + 6 \\ &= 104 \text{ dBp} \end{aligned}$$

Fig. 5.5

An approximate figure for the power level of a fan is

$$\begin{aligned} WL &= 90 + 10 \log W + 10 \log P \\ &= 90 + 10 \log W + 10 \log P'^{-4} \end{aligned}$$

where

WL = fan sound power referred to 10^{-12} W
W = rated shaft horsepower of fan
P = static pressure across fan in inches of water
P' = static pressure across fan in centimeters of water

The equation refers to the total sound power across the entire audio spectrum. Most fans generate less power above 500 Hz than below this frequency. In general, for the frequency range from 63–4000 Hz, the decrease in sound power for most fans comes to 12 dB, or 2 dB per increasing octave.

The sound power levels referred to 10^{-12} W for centrifugal and for axial compressors are approximately

$$WL_c = 81 + 20 \log HP$$

$$WL_a = 76 + 20 \log HP$$

where HP = compressor horsepower.

The sound pressure level of an omnidirectional compressor at a distance from the unit and at a given frequency is approximately

$$\begin{aligned} SPL &= WL - 20 \log r - 0.7 \\ &= WL - 20 \log R - 11 \end{aligned}$$

where

r = distance in feet
R = distance in meters

5.2 VIBRATION ISOLATION OF AIR-CONDITIONING SYSTEM EQUIPMENT

In the case of large air-conditioning system equipment installations, whether on the roof or in the basement of a building, two types of disturbances generated by the machinery deserve effective control in planning.

One type is called solid-born (solid-generated) and is solid-borne (solid-transmitted). This mechanical vibration travels in concrete with ten times the speed of sound in air, and usually has much greater energy than air-born and air-borne noise. Manufacturers rarely give vibration level data for this type of equipment, and this investigator sometimes had to to examine an existing installation to obtain pertinent values for a project.

The second type of disturbance produced by a large air-conditioning system is the air-borne noise, which generally is also of a high level.

The two units mostly responsible for the disturbances discussed above are the compressors and the distribution fans.

Fig. 5.6 shows the velocity level characteristics measured at the feet of fans varying in power from 3-20 horsepower, and at the feet of centrifugal refrigerators varying in capacity from 720 megacalories/hour (240 tons of refrigeration) to 1950 megacalories/hour (650 tons of refrigeration), when both types of equipment are resiliently suspended (there would be no displacement and therefore no velocity at the feet of the machines if they were tied to a concrete slab of infinite mass). Note that the vibration levels of the various size fans and compressors do not change greatly. This is generally due to the fact that the larger units are built more sturdily than the smaller ones.

Air-born noise levels about compressors in the 100-ton or larger range are generally in the order of 100 dB-A at 5 ft (1.5 m), and those of distribution fans are similarly high when measured at the duct intake opening. Outside the fans the noise levels are lower in relation to the sound transmission loss characteristic of the metal walls of the fan.

For a wholly satisfactory installation of a large air-conditioning system on the roof or in the basement of a building, a floating concrete floor is necessary. On the roof, the resiliently supported floor fulfills the double purpose of preventing the radiated air-borne signals from penetrating the rooms directly below the roof, and of acting as an inertia block for absorbing the energy of the solid-born vibrations of the equipment. The former type of noise reduction is made possible by the elastically coupled double barrier—the floating concrete floor and the structural slab below it. Sound transmission reduction by mass alone follows a logarithmic law, according to which doubling the mass of a single homogeneous barrier allows only a 4.5 dB increase in its sound transmission loss, while two barriers, structurally isolated, with sound-absorbent material in the interspace, secures 10 dB more noise level reduction for the same surface density (mass per unit area) as a single wall.

The installation of the equipment on a floating concrete floor in the basement also assures the attenuation of the mechanical vibrations by the inertia block which the floating floor represents, while the walls and ceiling of the mechanical room are the second structurally isolated barrier for the air-borne noise. In the case where the builidng adjoins apartment houses and other residential domiciles, a basement installation provides horizontal sound attenuation too, whereas a roof installation does not unless special enclosures or parapets are employed on the roof to protect neighbors.

The reason why air-conditioning equipment installations in basements are not well liked by architects and building owners is that they deprive this location of parking space for cars. With new regulations requiring large areas for parking near the structure, such basement locations for machinery often demand two and three lower levels, with consequent

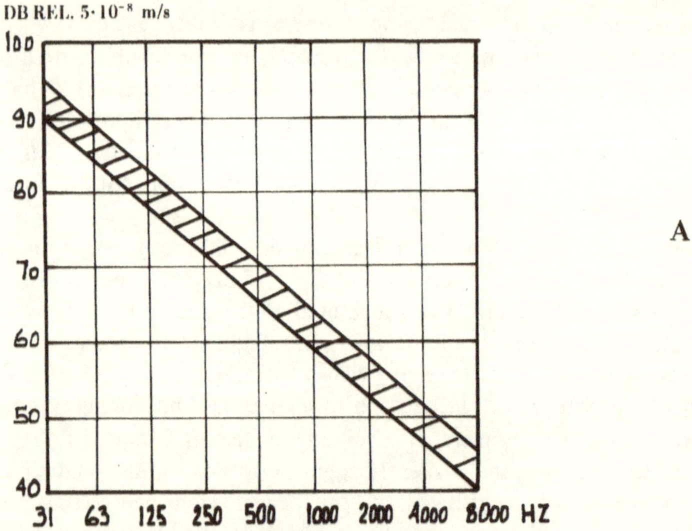

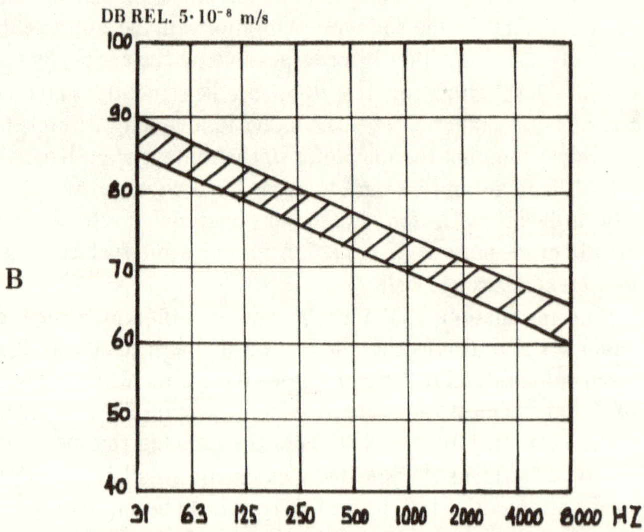

Fig. 5.6A Range of velocity levels with frequency for resiliently mounted centrifugal fans varying in capacity from 3 to 20 HP and in air volume flow rate from 5000 to 30,000 m^3/h. Measurement made at feet of machine.

Fig. 5.6B Range of velocity levels with frequency for resiliently mounted centrifugal refrigerators varying in capacity from 760 Mcal/h (240 tons) to 1950 Mcal/h (650 tons). Measurement made at feet of machine.

expense increase in construction. Hence a 400-ft^2 (37 m^2) mechanical equipment room would deprive the basement area by at least two car parking spaces.

When the external noise level of the equipment is 100 dB-A at 5 ft (1.5 m) and the internal noise level of a room should be *PNC* (Preferred Noise Criterion) 30 dB, equivalent to 39 dB-A the overall air-borne noise level reduction of the intervening barriers should be 100 – 39 = 61 dB.

The vibration attenuation requirements are in the order of 40 dB, as illustrated in Fig. 5.7 where the velocity level characteristic of a large system is represented by a solid line, which essentially follows the data shown before. The dashed line is the air-particle velocity level characteristic corresponding to *PNC* 30, which is equivalent to 39 dB-A. The difference between the two spectra represents the vibration isolation characteristic for the solid-born and solid-borne disturbances. Since the velocity level is referred to 5×10^{-8} m/s, the sound pressure levels pertain to the standard reference level of 2.0×10^{-5} N/m^2 or 2.0×10^{-4} μbar, according to

$$\begin{aligned} P_O &= \rho c V_O \\ &= 1.185 \times 344 \times 5 \times 10^{-8} \\ &= 0.00002 \text{ N/m}^2 \end{aligned}$$

where

ρ = density of air
= 1.185 kg/m^3
c = velocity of sound in air
= 344 m/s
P_O = reference sound pressure
V_O = reference air particle velocity

Such large mechanical vibration amplitude attenuations cannot be had by a resilient machine support only, since a 15-dB level reduction is just about maximum at the low frequencies. Hence an inertia block or floating concrete floor is also necessary for the air-conditioning system machinery

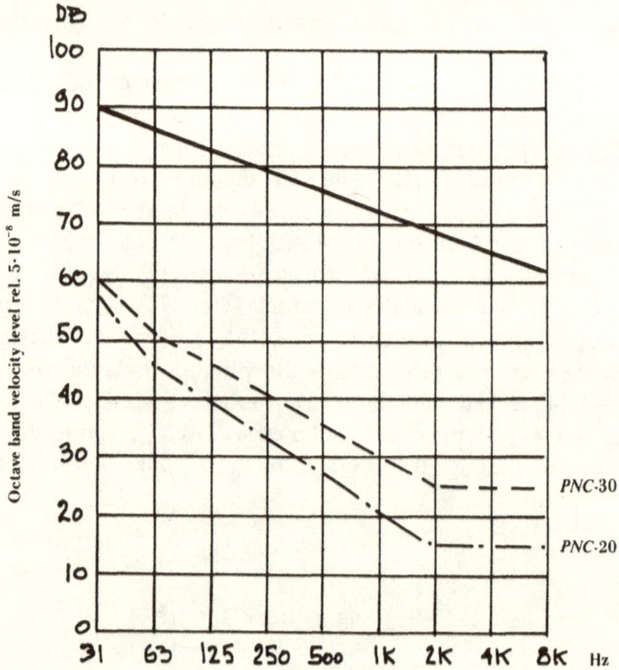

Fig. 5.7 Solid line represents velocity level characteristic of resiliently mounted centrifugal compressors between 240 and 650 tons of refrigeration. For a *PNC*-30 in adjoining room, a velocity level reduction of 37 dB at 125 Hz is required by intervening isolators.

of mass m. When the concrete mass M is 4 times that of the machinery, another 14 dB of vibration level reduction is achieved according to $20 \log(1 + M/m) = 20 \log(1 + 4) = 14$ dB. The remainder of the required vibration level reduction is generally obtained by distance and impedance mismatches between the various media through which the disturbance travels. For extremely low noise levels in the adjoining enclosures of a mechanical equipment room it becomes necessary to employ pneumatic isolators instead of springs, since such "air bags" can provide very low resonance frequencies for the mechanical system which such an installation represents. Indeed, a resonance frequency as low as 1 Hz can be had, which would normally require a 10-in. (25 cm) static deflection on the part of a spring. When the static deflection equals, e.g., 25 percent compression, the height of the spring comes to 40 in. (100 cm)—an impractically tall isolator!

When the roof of a building is considerably larger than the air-conditioning machinery, it may be desirable to employ there a resiliently supported concrete platform for the equipment, with sufficiently high sides or walls to form an enclosure. The vertical barriers, when they are at least as high as the machinery, act as sound screens whereby the lateral radiations from the source are either absorbed by the inside sound-absorbent treatment of the enclosure walls, or are sent skyward, or both. Fig. 5.8 shows an illustration of such an enclosure.

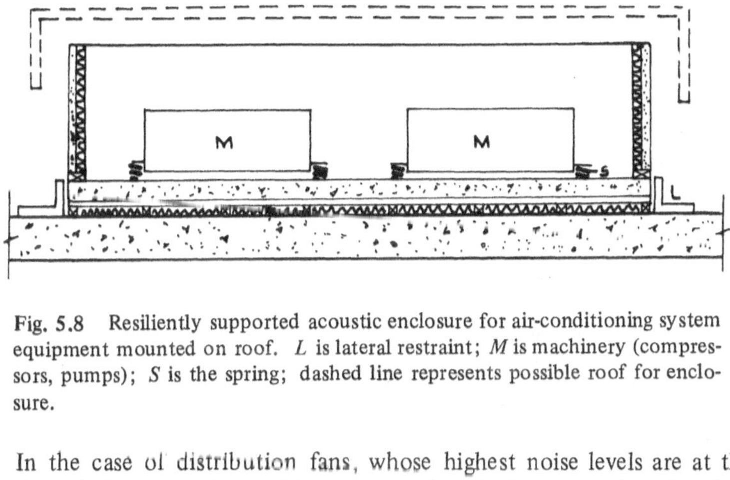

Fig. 5.8 Resiliently supported acoustic enclosure for air-conditioning system equipment mounted on roof. *L* is lateral restraint; *M* is machinery (compressors, pumps); *S* is the spring; dashed line represents possible roof for enclosure.

In the case of distribution fans, whose highest noise levels are at the intake of the main duct, either a convoluted plenum, a lengthy duct silencer (section of duct with noise traps), or a long sound-absorbently treated duct is necessary to lower the noise level, at the diffuser end of the duct, to a value recommended for the type room which is to be air-conditioned.

When it is not possible to obtain noise and vibration level data from the manufacturer, it is desirable to make measurements of an existing air-conditioning system installation of the type planned for the project under consideration. Even when no elaborate tests of the equipment are possible, a simple sound level meter reading with the A and C scales of ten provides highly pertinent data. Fig. 5.9 gives the approximate spectra of machinery obtained with only an A-weighted and a C-weighted reading. For example, when the difference between the C-scale and the A-scale reading of the sound level meter is 10 dB, the spectrum of the noise is that shown by the dot-dash line marked "10."

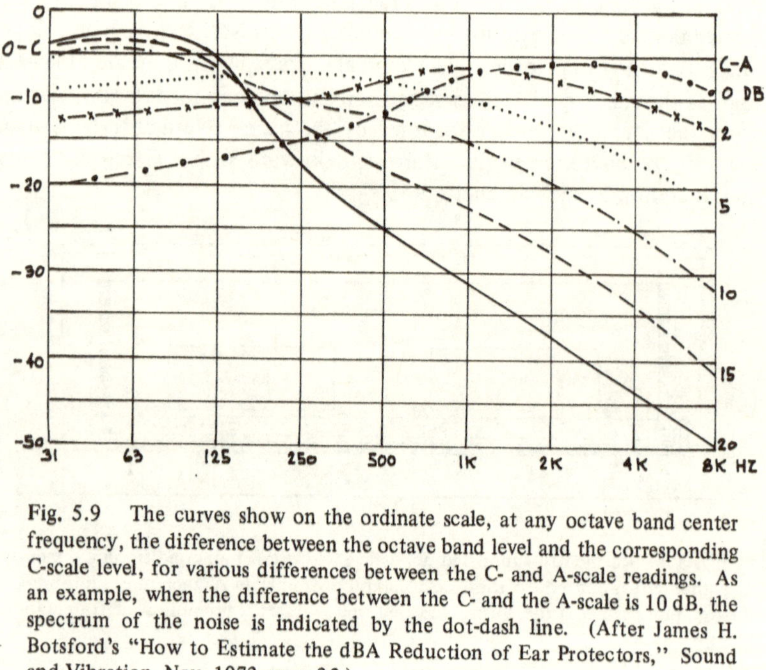

Fig. 5.9 The curves show on the ordinate scale, at any octave band center frequency, the difference between the octave band level and the corresponding C-scale level, for various differences between the C- and A-scale readings. As an example, when the difference between the C- and the A-scale is 10 dB, the spectrum of the noise is indicated by the dot-dash line. (After James H. Botsford's "How to Estimate the dBA Reduction of Ear Protectors," Sound and Vibration, Nov. 1973, page 32.)

To obtain vibration levels expressed in one reference quantity from vibration levels expressed in another reference quantity, use

$$L_D = L_V + 20 \log \frac{V_o}{D_o} - 20 \log f - 16$$

$$= L_A + 20 \log \frac{A_o}{D_o} - 40 \log f - 32$$

$$L_V = L_A + 20 \log \frac{A_o}{V_o} - 20 \log f - 16$$

$$= L_D - 20 \log \frac{V_o}{D_o} + 20 \log f + 16$$

$$L_A = L_V + 20 \log \frac{V_o}{A_o} + 20 \log f + 16$$

$$= L_D - 20 \log \frac{A_o}{D_o} + 40 \log f + 32$$

where L_D, L_V, and L_A are the displacement, velocity, and acceleration levels, and D_o, V_o, and A_o are the displacement, velocity, and acceleration reference quantities, and f is the frequency.

There are no international or even national standard quantities for reference displacement, velocity or acceleration. Many investigators use either 10^{-11} or 8×10^{-12} m for the displacement reference; 5×10^{-8} or 10^{-8} m/s for the velocity reference, and 10^{-3}, 10^{-5}, or 3.14×10^{-4} m/s^2 for the acceleration reference. Also

$$D = \frac{V}{2\pi f} = \frac{A}{4\pi^2 f^2}$$

$$V = 2\pi f D = \frac{A}{2\pi f}$$

$$A = 2\pi f V = 4\pi^2 f^2 D$$

To secure stability for the machinery isolated by springs, it is desirable to maintain a certain relationship between the unsprung height H of the spring; its working height h, that is, its height after the application of the load; its static deflection h, that is, the amount of compression after application of the load, and the spring diameter D. The table below gives these relationships in terms of the parameters described above:

	d	D	H	h
d		$0.33D$	$0.25H$	$0.33h$
D	$3d$		$0.75H$	h
H	$4d$	$1.333D$		$1.33h$
h	$3d$	D	$0.75H$	

As an example, assume that the static deflection of the spring should be 2 in. (5 cm) to achieve a resonance frequency $f_0 = 3.13/2^{1/2} = 2.21$ Hz. For such an isolator, the working height of the spring should be $h = 3d =$ 6 in. (15 cm); the unsprung height $H = 4d = 8$ in. (20 cm); and the coil diameter $D = 3d = 6$ in. (15 cm).

Another way to specify satisfactory conditions for the isolation scheme is to call for a spring whose lateral stiffness is equal to its axial stiffness; also the working height h of the spring should not be less than $2d$, to avoid damage to the spring when overcompression occurs.

When springs are subject to a deleterious weather effect, it is desirable to call for stainless steel springs, and when they are subject to possible high lateral forces, snubbers or lateral restraints should be incorporated in the isolation scheme.

5.3 SPECIFICATIONS AND CALCULATIONS FOR AIR-CONDITIONING SYSTEM NOISE LEVEL LIMITS

An air-conditioning system in a building is a complicated assemblage of components whose overall performance should be specified quantitatively. In most cases the mechanical engineer is well qualified to provide the number of required air changes for a room of given purpose and the desired air temperature differential between the outside and the inside of the structure. In respect to the maximum noise level of a room it is a mistake to say that it should be "non-annoyingly or barely audible," because such specifications are subject to personal interpretations and the hearing acuity of the listener. It is far better to state:

> The noise in the room, due entirely to the air-conditioning system, should not exceed *PNC* (Preferred Noise Criterion) 25 or 35 dB-A (or whatever maximum noise level the enclosure can tolerate) at all significant listening or sound-recording positions in the room, when it is empty and the windows and doors are closed.

The following pages are intended to provide basic design information for air-conditioning systems to be installed in noise-sensitive enclosures. It is important here to differentiate between a sound power level and a sound pressure level, since both are expressed in decibels. The former quantity, however, is generally designated in dBp, where *p* pertains to a reference level of 1 pW, while the latter quantity carries no such postscript. The sound power level of a source at a given frequency may simply be described as, for example, 90 dBp; to describe its sound pressure level it is also necessary to state the distance, as 90 dB at 50 ft (15.24 m).

Fig. 5.10 shows the sound attenuation characteristics of variously sized ducts lined with 1 in. (2.5 cm) thick fiberglass board of 3 lb/ft^3 (48 kg/m^3) density. It is seen that the common 12 in. × 12 in. (30 cm × 30 cm) square duct offers attenuation of only 0.7 dB/ft at 250 Hz and one of 2.1 dB/ft at 500 Hz. Thus a 5-ft long duct attenuates the octave-band whose mid-frequency is 250 Hz by only 7 dB while the 500-Hz signal is attenuated 21 dB.

Fig. 5.11 provides similar information when the cross-sectional area of the duct and the absorptivity characteristic of the duct liner are known.

Unlined ducts, regardless of size, rarely exhibit sound attenuation in excess of 0.5 dB/ft (1.5 dB/m) and frequently can provide no more than 0.25 dB/ft (0.75 dB/m). This is because the sound loss must be produced entirely by the vibrations of the duct walls. Thus so-called speaking tubes are used aboard a ship, where one person speaks into the tube at one end and another listens to the speech at the other end, with little loss in signal level.

So-called integral ducts show no more sound attenuation per unit duct length than lined sheet metal ducts. Because an integral duct is essentially a molded fiberglass conduit, with 1 in. (2.5 cm) thick walls, and externally covered with a sheet of aluminum foil, it cannot reduce the transmission of the noise inside the duct to the outside as much as a sheet metal duct can with its much heavier external "skin." When this noise transmission to the outside of the duct is not of much significance, as in an attic with a heavy ceiling, an integral duct generally is cheaper than a lined sheet metal one, and performs as well.

Calculating the required sound attenuation for the ducts in an air-conditioning system may be more or less difficult, depending on the complexity of the installation. It is relatively simple when the investigator knows the volume and purpose of the room which is to be air-conditioned, the air-volume in the room which is to be changed a specified number of times

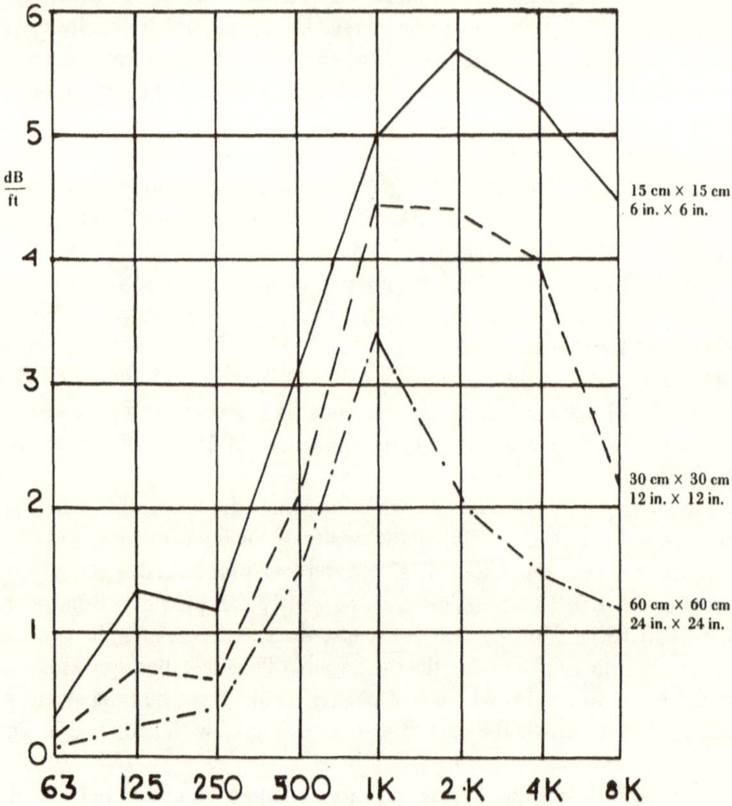

Fig. 5.10 Sound attenuation in decibels per foot of air-conditioning system duct lined with 1 in. thick fiberglass board, as a function of octave-band center frequency. Multiply dB/ft by 3.28 to obtain dB/M.

per hour, the available length for the duct runs, and the sound power level characteristic of the distribution fan. It is not a simple manner to calculate the sound attenuation in the ducts when several rooms of various volume and purpose are to receive climatic control by a single fan blowing air into a main duct from which side branches have to deliver specific air volumes. The example below is meant to illustrate the basic method for computing

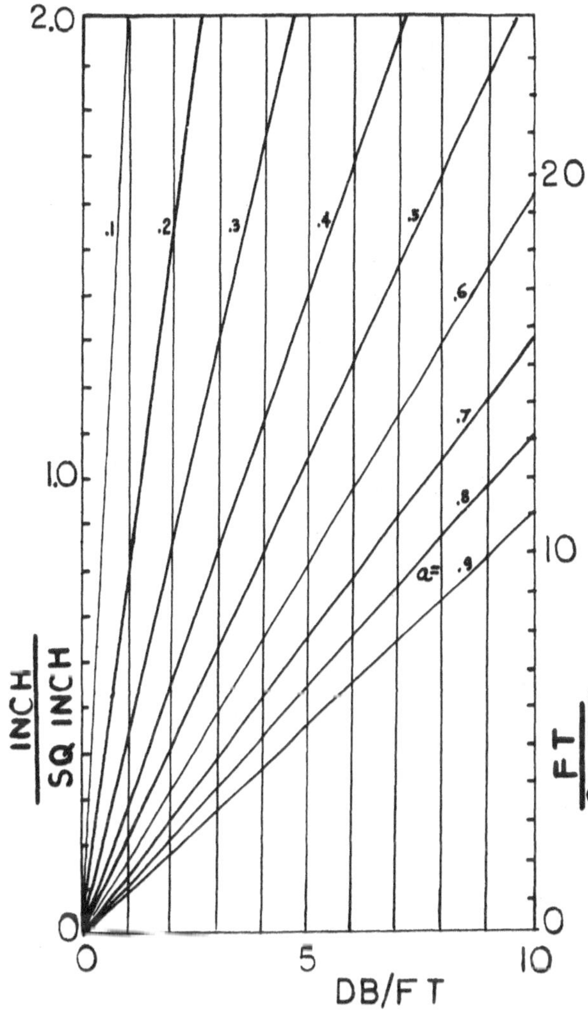

Fig. 5.11 Chart for calculating the sound attenuation characteristic of a duct of known cross section lined with a sound-absorbent material of known absorption characteristic. Multiply the ratio of duct perimeter in cm to the duct cross section in cm^2 by 2.54 to obtain the corresponding ratio of duct perimeter in inches to the duct cross section in square inches. Similarly, multiply the ratio of duct perimeter in m to the duct cross section in m^2 by 0.0929 to obtain the corresponding ratio of duct perimeter in feet to the duct cross section in square feet. Multiply dB/ft by 3.28 to obtain dB/m.

the required duct attenuation when the chief concern of the system is to air-condition one large enclosure.

Assume a sound-recording studio 20 ft × 30 ft × 50 ft (6.1 m × 9.1 m × 15.2 m), with a volume of 30,000 ft^3 (849 m^3), and a reverberation time of 0.78 s at 1000 Hz, is to have a noise level characteristic of *PNC* (Preferred Noise Criterion) 25, with a maximum noise level of 25 dB at 1000 Hz. There are to be 12 air changes per hour. The total volume change is $3.0 \times 10^4 \times 12 = 3.6 \times 10^5$ ft^3 (10188 m^3) per hour or $3.6 \times 10^5/60 = 6000$ ft^3 (169 m^3)/min. If the air speed should be no more than 750 ft (228 m)/min to avoid turbulency at the diffuser, the cross-sectional area of a single duct comes to 6000/750 = 8 ft^2 (0.74 m^2), and of a double duct the cross-sectional area is half as much, or 4 ft^2 (0.37 m^2) equal to a 24 in. × 24 in. (60 cm × 60 cm) duct.

The power level required at the diffuser of the room for a maximum sound pressure level of 25 dB at 1000 Hz and one of 43 dB at 125 Hz to maintain *PNC*-25 in the reverberant part of the enclosure is given by

$$\begin{aligned} WL &= SPL + 10\log V - 10\log T - 29.3 \\ &= SPL + 10\log V' - 10\log T - 14 \end{aligned}$$

where

WL = sound power level relative to 1 pW
SPL = sound pressure level relative to 2.0×10^{-4} μbar
V = volume of room (ft^3)
= 30,000 ft^3
V' = volume of room (m^3)
= 849 m^3
T = reverberation time
= 0.75 s at both 125 and 1000 Hz

The required power level, therefore, is

$$\begin{aligned} WL_{1000} &= 25 + 10\log 30{,}000 - 10\log 0.75 - 29.3 \\ &= 41.7 \text{ dBp at 1000 Hz} \\ WL_{125} &= 43 + 10\log 30{,}000 - 10\log 0.75 - 29.3 \\ &= 59.7 \text{ dBp at 125 Hz} \end{aligned}$$

Assume that *PNC*-25 is to be maintained not only in the reverberant part of the studio, but also at a distance from the diffuser where the direct sound energy density exceeds the reflected sound energy density, e.g., at 5 ft (1.5 m) from the diffuser. The pertinent equation is

$$\begin{aligned} WL &= SPL + 20 \log d - 10 \log Q + 0.7 \\ &= SPL + 20 \log d' - 10 \log Q + 11 \end{aligned}$$

where

d = distance from diffuser (ft)
d' = distance from diffuser (m)
Q = directivity factor of the source
= 2 for the diffuser flush with the ceiling

Hence, for 1000 Hz,

$$\begin{aligned} WL_{1000} &= 25 + 20 \log 5 - 10 \log 2 + 0.7 \\ &= 39.7 \text{ dBp} \\ WL_{125} &= 43 + 20 \log 5 - 10 \log 2 + 0.7 \\ &= 54.7 \text{ dBp} \end{aligned}$$

Chapter 6

Sound-Retardant Windows and Doors

6.1 SINGLE-PANE WINDOWS

For a single-pane window, the two most significant factors regarding its sound insulation are the surface density of the single sheet of glass, that is, its mass per unit area, and the composition or fabrication of the pane. Secondary factors are the seals around the perimeter of a light, and the damping material under the pane which acts as a vibration attenuator at the coincidence frequency of the pane when the wavelength of the incident sound equals the wavelength of the free bending wave in the pane.

Fig. 6.1 shows the variation of the sound transmission loss of a single pane of glass as a function of its thickness in inches for various frequencies, when the glass is not laminated (the second significant factor noted above). When it is laminated, as in the case of "Acousta-Pane," safety glass, etc., it is safe to add 2 decibels to the mean transmission loss, and 3-4 decibels when the transparent plastic layer between the two sheets of glass is thicker than 1/32 in. (0.08 cm).

To provide a good seal for the sheet of glass, a plastic, nonhardening preparation instead of putty should be used. There are also Neoprene extrusions available for the purpose, and so-called "lock-and-key" fasteners as shown in Fig. 6.2. The seal should extend into the glass pane at least ½ in. (1.27 cm).

6.2 MULTI-PANE WINDOWS

Fig. 6.3 shows details of construction for a double-pane window. For the same total glass thickness of a double-pane window that a single-pane one has, the sound transmission loss of the double-pane unit may be from 4–10 decibels greater, depending on the distance of separation of the sheets

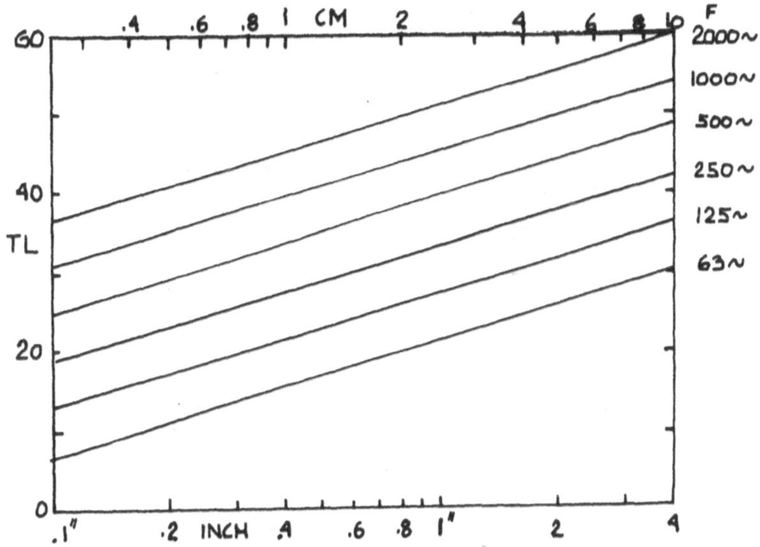

Fig. 6.1 Variation of sound transmission loss *TL* with thickness of the pane at various frequencies, assuming no leakage losses.

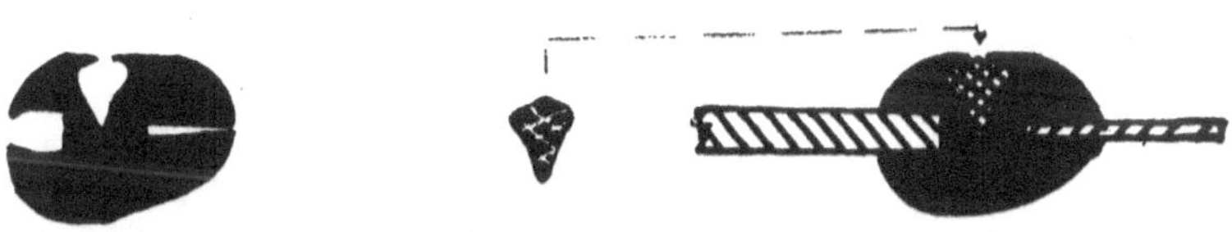

Fig. 6.2 Neoprene lock-and-key extrusion for resilient support of sound-retardant window pane

of glass. Surprisingly, a triple-pane window with the same total glass thickness as a double-pane window has a sound transmission loss only 1-3 decibels greater.

Instead of using the cork-wood-frame construction shown in Fig. 6.3, a Neoprene extrusion may be employed, and a ½ in. (1.27 cm) thick cork or felt pad may be employed under the Neoprene surround for extra cushioning and structural isolation between panes.

A ½ in. or 1 in. thick sound-absorbent material should be placed on the walls between the panes, so that the sound, after having passed through

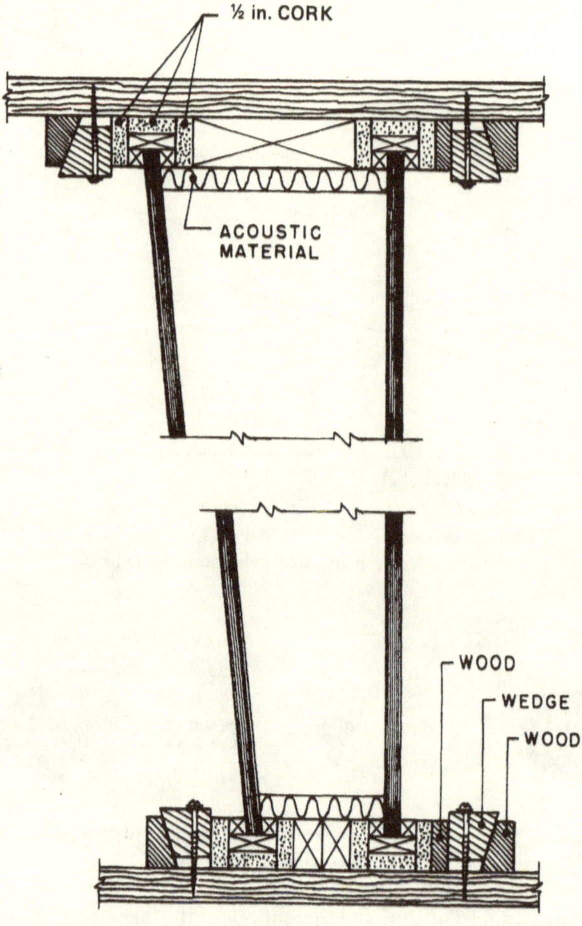

Fig. 6.3 Details of construction for a double-pane sound-retardant window.

one sheet of glass, will to some extent become absorbed in the interspace before further transmission.

Besides the system of double-pane construction outlined in Fig. 6.3, there are also systems which employ lock-and-key extrusions, H-type Neoprene gaskets, and PVC reglets with spline gaskets manufactured by the

Standards Products Co. in Port Clinton, Ohio.*

Fig. 6.4 shows details of construction of a double-pane window using PVC reglet.

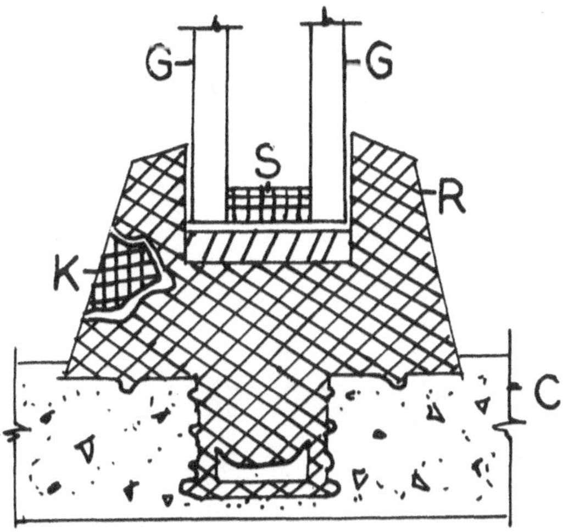

Fig. 6.4 PVC reglet cast into concrete window sill. R: reglet; S: spacer; G: glass panes; K: key for closing reglet; C: concrete sill

The rubber extrusions also act as a damping material for the sheets of glass when they vibrate strongly at either their coincidence frequency or their panel resonance (when their mass reactance equals their stiffness reactance).

Since it is the projected wavelength of the incident sound on the panel which determines the coincidence effect, the resonance frequency is dependent on the angle of incidence, as shown in Fig. 6.5.

Fig. 6.6 shows the variation of the sound transmission loss of a double-pane window as a function of glass thickness and pane separation. The transmission losses are average figures, dependent only on the surface density of the window and the pane separations, and do not include resonance effects.

To avoid equal coincidence frequencies and mass-stiffness resonance frequencies in a double-pane window, the thicknesses of the two panes

* Distributed in California by Marketing Management Inc. 11656 Vanowen Street, North Hollywood, 91605 (Tel: 213-982-4162).

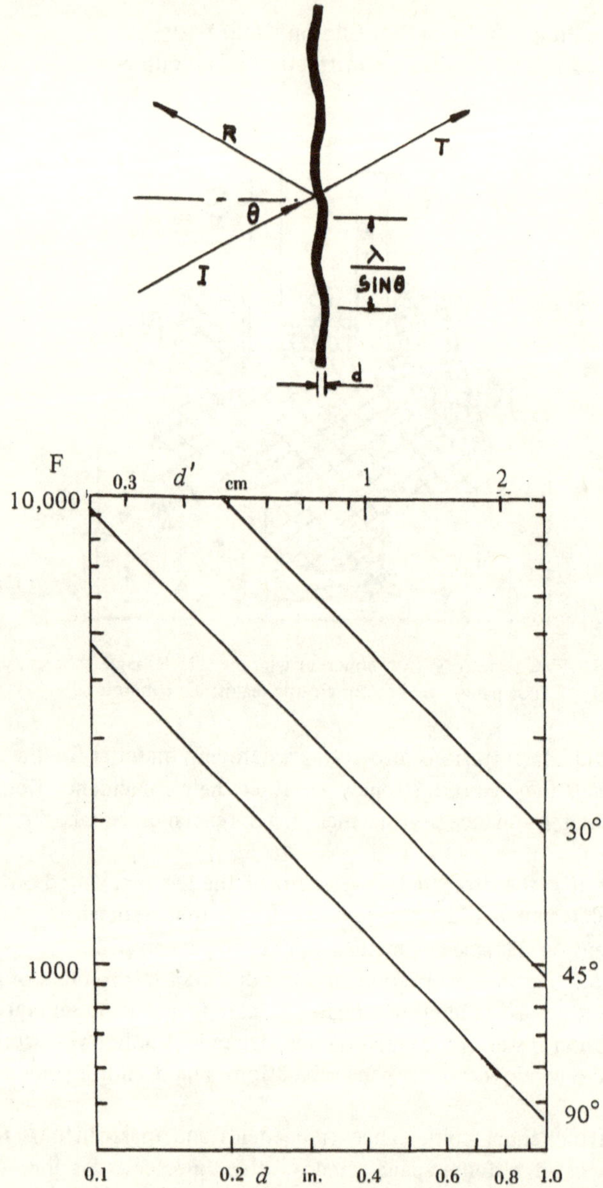

Fig. 6.5 Top: Coincidence effect. Bottom: Variation of coincidence frequency with glass thickness for various angles of sound incidence

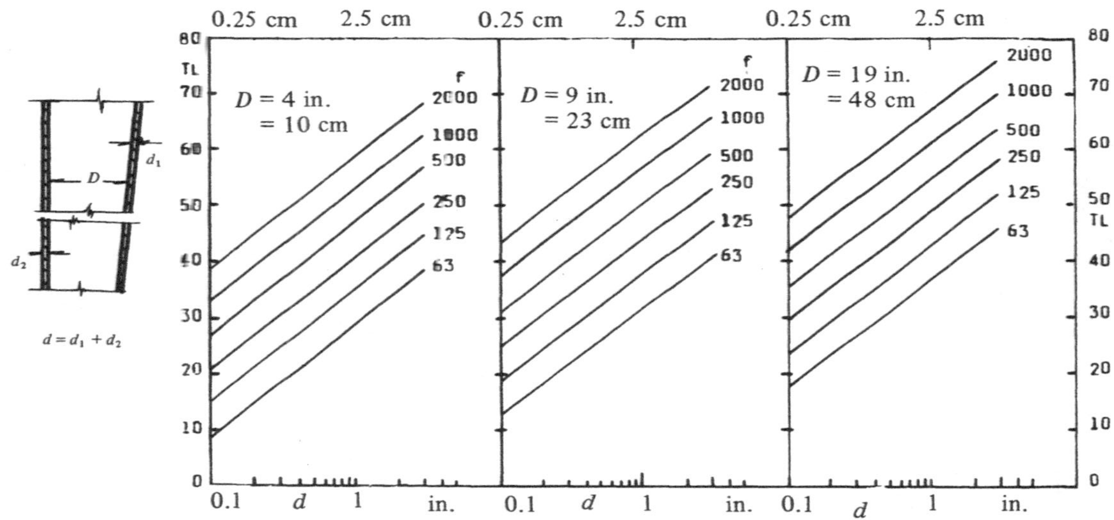

Fig. 6.6 Variation of sound transmission loss of double-pan window with total glass thickness at various frequencies and for various pane separations

should not be the same. There appears to be no optimal ratio for the two thicknesses, although in practice a 2:3 ratio is often used. Thus, when one pane is ½ in. thick, the other should be ¾ in.

The vertical angle of the panes should not be less than 6 degrees in respect to each other, to avoid a strong standing wave between the sheets of glass when a prolonged note is incident on the window.

As stated previously, laminated glass sheets exhibit a somewhat higher sound transmission loss than a homogeneous sheet, but there is a slight loss in light transparency for laminated panes. Hence, such sheets are unacceptable for projection ports in theaters, review rooms, etc. For maximum luminous transmittance Lucite AR sheets are sometimes used for projection ports.

For a composite barrier, like a brick or stucco wall with a window, the total sound insulation is not that of the window, even though it may weigh much less per square foot than the wall. The total sound transmission loss at any one frequency is a complicated factor which involves not only the surface densities of the two materials, but also their relative areas.

Mathematically the total sound attenuation of a composite barrier is given by

$$TL = TL_0 - 10 \log \left[(1 - i) + i \times 10^{(TL_0 - TL_1)/10} \right]$$

If $i = 0.2$

$$\begin{aligned} TL &= TL_0 - 10 \log \left[(1 - 0.2) + 0.2 \times 10^{(TL_0 - TL_1)/10} \right] \\ &= TL_0 - 10 \log \left(0.8 + 0.2 \times 10^{\Delta TL/10} \right) \end{aligned}$$

where

TL_0 = sound transmission loss of wall at frequency f
TL_1 = sound transmission loss of window at frequency f
ΔTL = $TL_0 - TL_1$
i = percentage of window area

Fig. 6.7 is a graphic illustration of the above equation.

As an example, consider a control room window with a sound transmission loss $TL_1 = 45$ dB at 500 Hz. It occupies 50 percent of the partition between control room and associated studio, with the partition having a transmission loss of $TL_0 = 55$ dB at 500 Hz. For this condition, $TL_0 - TL_1 = 55 - 45 = 10$ dB. Enter the ordinate of the figure at the 10 dB

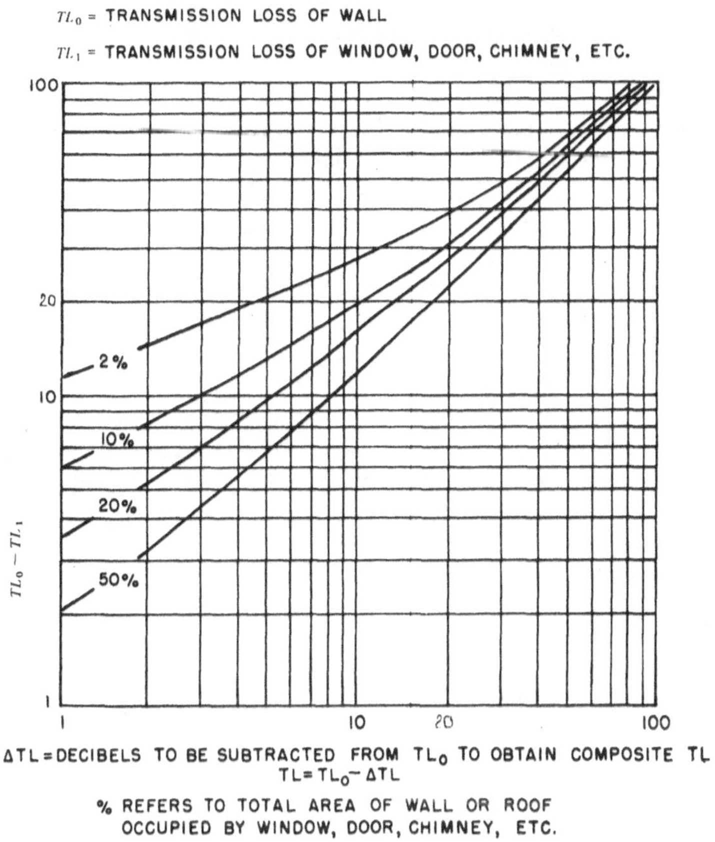

Fig. 6.7 ΔTL = decibels to be subtracted from TL_0 to obtain composite $TL = TL_0 - \Delta TL$; % refers to total area of wall or roof occupied by window, door, chimney, etc.

point and move horizontally on the line to the 50 percent curve, and from there downward to the ΔTL abscissa to the point 8 dB. Hence, the composite TL of the partition with its window comes to $TL_0 - \Delta TL = 55 - 8 = 47$ dB at 500 Hz.

6.3 SOUND-RETARDANT DOORS

The sound insulation of a door is not unlike that of any solid panel in that it must follow the acoustic mass law, that is, its insulation is a logarithmic function of its mass per unit area.

Doors often exhibit considerable sound leaks at their perimeters. A 3 ft × 7 ft (0.9 m × 2.1 m) door has a perimeter of 20 ft (6 m); even when there is only 1/16 in. (0.15 cm) leak around the perimeter, the leakage path comes to more than 1 $in.^2$ (6.45 cm^2)–a large hole for the sound to pass through. For this reason, sound-retardant door manufacturers often offer rabbeted or grooved frames with their doors to assure a good seal.

Two medium-density doors of, e.g., 5 lb/ft^2 (24 kg/m^2) at the entrance to a noise-sensitive enclosure, generally offer more noise reduction than a single door weighing as much as 20 lb/ft^2 (27 kg/m^2).

Doors like other wood panels exhibit a variety of resonances. To reduce the ill effects of the vibrations, in which the transmission loss of the unit becomes small, doors are often built up in layers of wood and a resistive element like lead or a viscoelastic material to dampen the large panel excursions at these narrow frequency regions.

As would be expected, doors should either butt up against a threshold, or they should have a drop-bottom seal which closes the gap between the floor and the lower edge of the door when it is closed. Door stoppers which press the door against the door frame are another desirable element in sound-retardant door construction.

Chapter 7

Instrument Isolation for Multiple Track Recording

For reasons of greater sales appeal it has become practice in the recording industry to sell records and magnetic tapes whose stereophonic or quadraphonic sound tracks were secured by mixing down as many as 32 tracks, while for motion pictures as many as 64 tracks have been used. When it is not possible to achieve clean tracks for each instrument or band section in one multiple track recording session, two or more sessions may be called, each devoted to a different type of instrument (woodwinds, percussion, strings, etc.). This is generally the case when the studio is relatively small or rather live so that it is not possible to space the performers far enough apart to acquire tracks not blemished by signals from adjoining sources. So that the musicians of the later sessions may stay in time with the previously recorded music, they must listen to it over earphones.

To accommodate as many instruments as possible in a studio, flats or sound-shielding panels are often placed between performers. In some studios, retractable partitions are pulled from the wall to achieve similar results. In some instances, cells or small portable rooms are introduced in which a piano, drum, organ, etc. are placed. Booths are also provided for vocalists and choruses, particularly when the song is intended to be released both with and without the accompanying music.

Fig. 7.1 shows a flat, 8 ft (2.5 m) high, which separates a flutist from a trombone player. Each player is 5 ft (1.5 m) from the barrier, which is assumed to have a sufficiently great surface density to allow a negligible amount of sound to pass directly through it. Also, it is assumed that the players are deep enough in this divided space so that sound transmission around the leading edge of the barrier is negligibly small. Also, transmission under the barrier is not assumed possible.

For such an arrangement, the determinative factor for the sound isolation between the two players is the path difference between the diffracted sound (14 ft or 4.3 m) and the direct sound which would pass between the

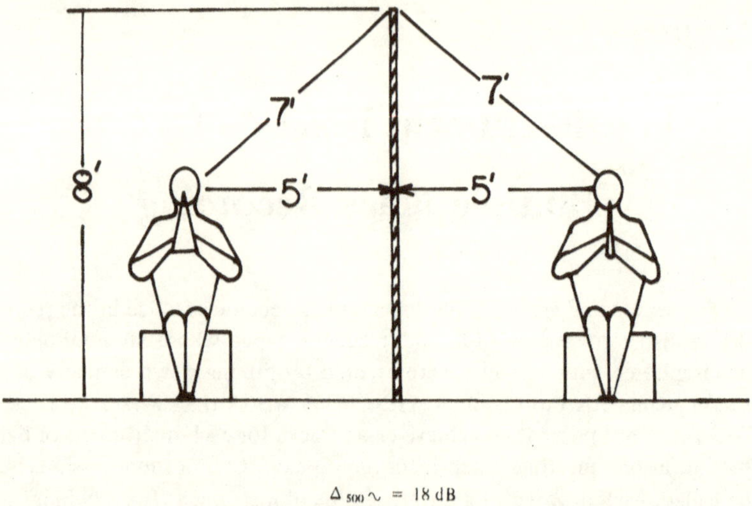

Fig. 7.1 An 18 dB excess sound attenuation at 500 Hz is provided by a 8-ft (2.4 m) high "flat" between two instrumentalists 5 ft (1.5 m) from the barrier.

two instrumentalists if no barrier were between them (10 ft or 3 m). For a path difference $d = 14 - 10 = 4$ ft (1.3 m), the excess sound attenuation at 500 Hz comes to 18 dB. Excess sound attenuation refers to the sound pressure level reduction over and above that due to spherical divergence only when no sound shield is between the two sources. As will be explained later, the excess sound attenuation is a function of the frequency of the signal which passes over the top edge of the obstacle.

When the barrier between two such players is not a single screen but a large obstacle, the path difference may be calculated by the method shown in Fig. 7.2. Here, the performers are separated by a vocal booth, and are again assumed to be 5 ft (1.5 m) from the sound interceptor. The path difference between the diffracted signal across the top of the booth and the direct sound between the players in the absence of the enclosure comes to 5 ft (1.5 m) and the excess sound attenuation at 500 Hz is calculated to be 19 dB.

The above sound pressure level reductions are based on the condition that the studio has a sufficiently short reverberation period so that signals reflected from the room boundaries do not contribute appreciably to the

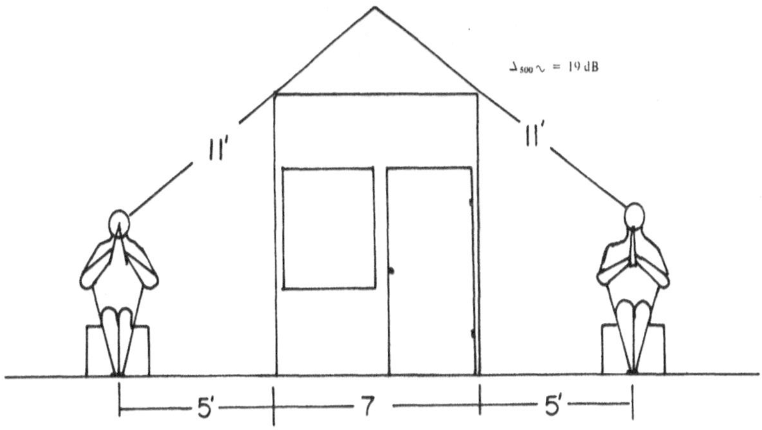

Fig. 7.2 When a 10-ft (3.1 m) high and 7-ft (2.1 m) wide vocal booth separates two instrumentalists as shown, the excess sound attenuation at 500 Hz is 19 dB.

sound level at the various microphones within the divided spaces. This requirement can be more easily fulfilled when the microphones have a cardiod directional response. When, as in Fig. 7.1, the players are not sufficiently deep in their cells so that sound transmission can also take place around the leading edge of the barrier, the excess sound attenuation will be less. Thus, when the path length around this vertical edge is equal to the path shown across the top of the shield, the sound pressure level at the microphone will be 3 decibels greater, and the sound attenuation will be only 15 dB at 500 Hz.

Fig. 7.3 illustrates the isolation of two performers by means of an extensible partition. Here the greatest sound transmission occurs around the leading vertical edge of the barrier, since it extends from floor to ceiling. Also, it is assumed that negligible sound transfer takes place through the wall pocket into which the accordion-fold type partition is pushed when not in use as a baffle. The arrangement pictured provides a path difference of 14 ft (4.3 m) and an excess sound attenuation of 24 dB at 500 Hz. Again it must be assumed that the sound can not pass to an appreciable extent through the partition directly.

Fig. 7.4 shows a space divider which is hinged to the wall of the studio. The 7-ft (2.1 m) long and 8 ft (2.4 m) high panels are reflective on one side and sound-absorbent on the other. By properly orienting them when

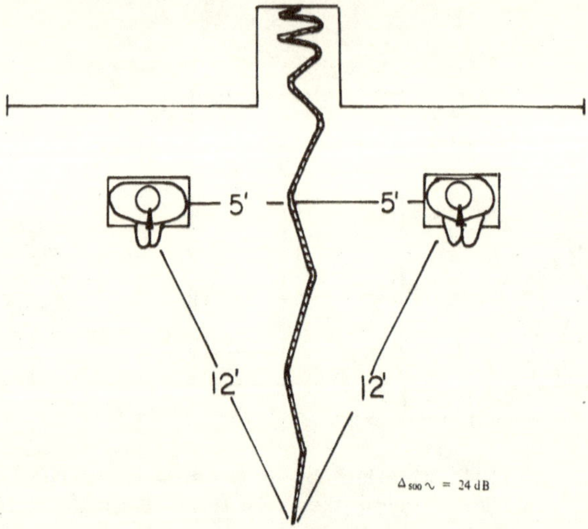

Fig. 7.3 When the path difference between instrumentalists is 14 ft (4.3 m), the excess sound attenuation at 500 Hz is 24 dB.

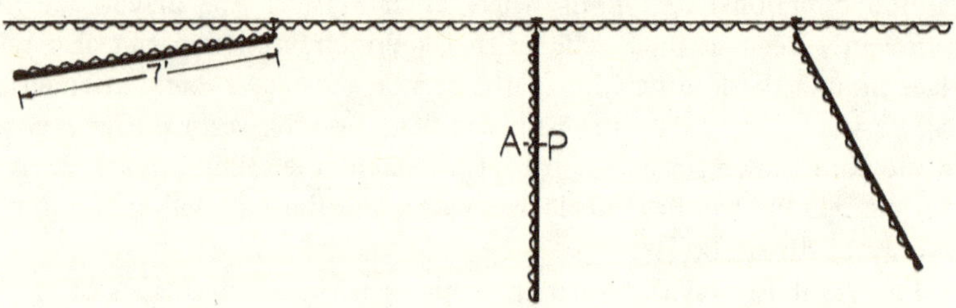

Fig. 7.4 Hinged 7-ft (2.1 m) wide and 8-ft (2.4 m) high panels, reflective on one side and sound-absorbent on the other, may be used as space dividers between adjacent instrumental sections for multiple track recording. Some studios prefer the installation of a plastic window in the panels so that adjoining performers may see each other for better timing.

they are installed on two opposite sides of the studio, these walls can be made almost entirely reflective or absorbent; or else, extended, can be formed into instrument cells for multiple track recording. The panels should come to the floor as closely as possible, so as to reduce the sound transmission under their lower edges when they are employed as dividers between band sections.

An added feature to the hinged splays consists in the installation of plastic windows in the panels so that adjoining performers may see each other for better timing.

Splays or hinged panels of the type discussed above require a solid wall to fasten them to, unless casters are placed under their lower edges to support part of the load. The weight of a 8-ft (2.4 m) high and 7-ft (2.1 m) long panel constructed of ½ in. (1.25 cm) thick hardwood plywood, adequate framing for it, with 1 in. (2.5 cm) thick acoustic material applied to one side of the panel, comes close to 150 lb.

Extensible partitions may also be installed on the ceiling of a studio instead of on the walls, so that the "curtain" is pulled down instead of out (of the wall). This may be desirable when there is insufficient wall depth to hide the accordian-edge "flat" in the closed position, which may require a box or cabinet 2.5 or 3 ft (0.7 or 0.9 m) deep (often well hidden in a suspended acoustic ceiling).

Whatever sound interceptors are employed, they should be sufficiently attractive to meet the interior decor of the studio. Fir plywood and unpainted Masonite shields are to be discouraged even though they may be cheaper.

Fig. 7.5 represents a graph for the rapid determination of the excess sound attenuation when the path difference *d* is known for the set-up under consideration. It is seen that the excess sound attenuation is less for the bass than it is for the treble. The decrease in sound level reduction comes to 3 dB per lower octave. Thus for any of the 500-Hz attenuations shown on the figures, the attenuation for 250 Hz is 3 dB less and that for 125 Hz is 6 dB lower.

To secure maximum isolation between woodwind instruments within a minimum space, recordings have sometimes been made by placing the sound radiating part of the flutes, saxophones, clarinets, etc. in a wooden or metal box which also contains the recording microphone. Manipulation of the stops or valves on the instrument must be done by inserting the player's arm through a sound-absorptive sleeve in the box, as shown on Fig. 7.6. Some performers object strenuously to this type of recording, claiming that they cannot do justice to the music when they are so encumbered. Others claim that, even though they are not greatly annoyed by the playing set-up, the recorded music lacks in quality by introducing box resonances which are not ironed out or dampened by whatver acoustic treatment is installed in the instrument enclosure.

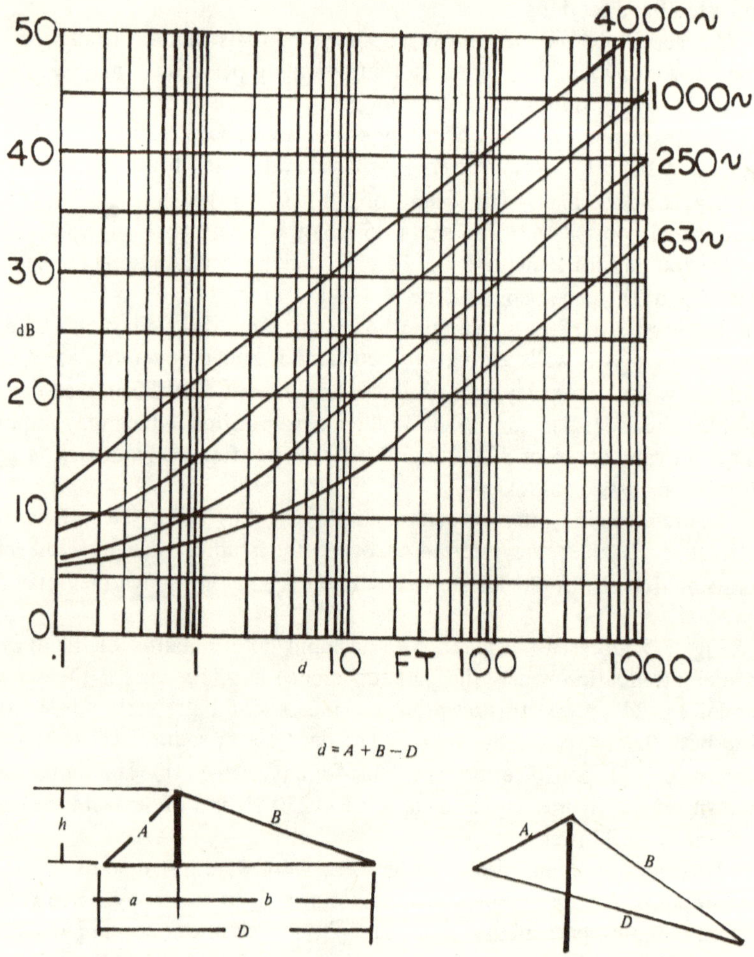

Fig. 7.5 Chart for the graphic evaluation of the excess sound attenuation of a barrier for different frequencies.

Complete sound isolation between instruments or band groups may be obtained by placing the different performers in separate rooms entirely, establishing visual contact with mixer, conductor, and other players by closed-circuit television.

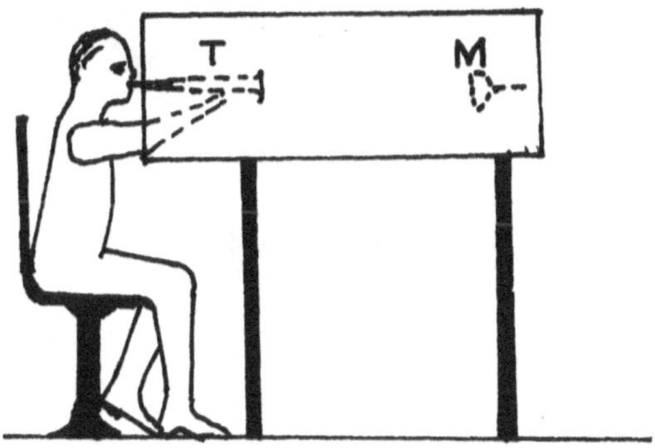

Fig. 7.6 Isolation of instruments by means of a sound-absorbently lined box; T: trumpet; M: microphone.

Chapter 8

Studio Testing

8.1 GENERAL

There is no single-number index for describing the acoustics of a room. One cannot measure the "frequency response of an enclosure," because it has many responses, depending on the position of the observation point in space and the signal arrival time at this point after generation by the source. Nor can one measure the degree of uniformity of the transient or steady-state sound-pressure distribution in the room, that is, its sound "diffusivity," which even lacks standard definition. Such often-used terms as "warmth" of a room, its "liveness," "clarity," "brilliance," etc., are equally nebulous quantities incapable of quantification.

This investigator, as a standard routine in studio testing, has developed the following measurements:

1. *Noise level.* The simplest test for noise in a room is the observation of a sound level meter set on the A-scale of the instrument. The air-conditioning system should be in full operation during the test, and all doors and windows should be closed.

A better assessment of the room acoustic disturbances consists in a 24-h noise level recording to learn of unexpected noises which might penetrate the enclosure during the night, early morning, etc.

2. *Vibration level.* A test similar to the above may be made, but with an accelerometer instead of a microphone as the signal pick-up device.

3. *Rattles.* The vibration of loose panels, lamp fixtures, etc., in the studio may be detected by generating in the room a high-amplitude frequency sweep, starting with as low a frequency as 20 Hz (if the loudspeaker system in the room is capable of reproducing so low a sinusoid), and cranking the oscillator very slowly through the entire audio frequency range.

4. *Reverberation time.* By this more or less standardized method of evaluating the reverberation time in the room one learns of its reverbera-

tion time characteristic, that is, its variation with frequency. Generally, by using 1/3 or 1/2 octave random noise signals, the reverberation time characteristic does not vary markedly from point to point in the studio. When discrete frequencies are employed, a slightly greater variation in the characteristic may be noticed, particularly at greater distances from the source in the room. The greatest variation occurs when so-called "glide tones" are generated, that is, closely spaced sinusoids. This is a method developed by the firm of Bruhl & Kjar, and special equipment is required, although with a sufficiently long paper roll in the high-speed graphic level recorder and sufficient patience on the part of the investigator who then has to make hundreds of reverberation time measurements at closely spaced signals, the same results may be obtained.

5. *Sound transmission between studio and control room.* This may be done by generating signals in one room and measuring the transmitted sound in the other. One should also position a microphone in the studio, turn the gain of the mixing console in the control wide open, and listen for any possible acoustic feedback if ever so low an acoustic signal is generated in the studio.

6. *Loudspeaker response.* In the control room the most significant position is that of the mixer or the positions of the mixers, if more than one operates the console. There are many methods of testing the loudspeaker response at the mixers's position, including time-delayed spectroscopy. This investigator prefers the use of two microphones, 7 in. apart, at the height of the mixer's ears at the console. The outputs of the two microphone preamplifiers should first be rectified before mixing them. This will avoid sharp "peaks" and "dips" in the response which may be due to phase differences in the microphones or amplifiers. Also, for a double check, the microphones should be interchanged.

The sound-pressure level produced by the monitor loudspeakers at the mixing console should be sufficiently high to include any anomolous signals which might be generated in the room by resonance vibrations of the walls, ceiling, or console panels. One may use either sinusoids or random noise signals passed through narrow band-pass filters in the loudspeaker amplifier.

Equalization of the system may then be made for "best" sound, often an entirely subjective quantity. In motion picture theaters and review rooms the acoustic response of the speakers in the auditorium is often made to correspond to the standard response specified by the SMPTE, the Society of Motion Picture and Television Engineers. In disk and tape recording studios a similar standardized response by other technical bodies is sometimes used, although in this investigator's experience, this response is not always employed, or if it is employed, it is "improved" upon by the technical director or chief mixer of the facility.

The Sennheiser Microphone Co. makes a plastic dummy head, complete with ear flaps and ridges, which has a small condenser microphone installed where the human ear is located, one on each side of the head, 7.25 in. (18.4 cm) apart. Similarly, an investigator may apply such microphones to his own head and direct their outputs to the proper instrumentation for measuring the "live" frequency response of monitor loudspeakers. The MKE2002 microphones are nearly omnidirectional when not installed on the dummy head, but become cardiodal in their directional pattern when applied to the "Kunstkopf."

7. *Acoustic Condition Factors.* There are a number of acoustic quantities generally described as "qualitative elements" which may be evaluated instrumentally. Thus, one may measure the sound energy which arrives at a listener's position after a given period, e.g., within 1/16th of a second, and compare it to either the total energy of the pulse or word, or compare it to the energy in the remainder of the signal. The "problem" with many of these acoustic condition factors is that there is no illustrious standard against which the results of the test may be compared. One may form a subjective opinion about the acoustics of a room which is not intimately connected with its reverberation characteristic, but little is known about the quantification of such aural impressions, and less about any ideal condition for listening in rooms. Does the ear prefer reflections from overhead to lateral reflections? To what extent is phase distortion important in recording a complex signal? To what extent is a nonexponential sound decay in a room detrimental to its listening conditions?

For decades inexperienced studio mixers and associated personnel have measured the sound level variation with frequency at distant parts from the sound source in a control room, and have wondered and worried about the results. Invariably the word "standing waves" enters discussions about the marked peaks and dips in signal amplitude at any such location, and the equally large variations of the acoustic response with microphone position. Panels are hung up and sound-absorbent treatment is installed at various surfaces in the enclosure to remove these "standing waves."

Every room however exhibits this type of sound interference pattern, including anechoic chambers, particularly at the low frequencies. It would be necessary to remove the normal modes in a closed space which is impossible because it is the nature of sound in a room.

For this reason it is desirable not to employ discrete frequencies for the signal, or even frequency-modulated sinusoids, but narrow bands of pink noise to learn of the preponderance of the high or low notes at a given listening position.

In control rooms, where the mixer is near the "critical distance," (where the direct sound energy equals the generally reflected one), the dips and peaks of the loudspeaker response should be much less violent. This is for the reason that the reflected sound can not produce the interference pattern as in the far field, and any change in level at all frequencies should approach the 6 dB change for double the distance between the microphone and the loudspeaker, or a somewhat smaller change because the transducer is leaving the near field.

8.2 ACOUSTIC RESPONSE OF RERECORDING STUDIO

To achieve a final sound track that can be satisfactorily reproduced in the theater, the electroacoustic system of the studio's rerecording chamber must be standardized. A detailed description of the situation was prepared by the Research Council of AMPTP (Assocition of Motion Picture and Television Producers) and printed in the January 1975 issue of the Journal of the SMPTE (Society of Motion Picture and Television Engineers).

Very briefly stated, the optic/acoustic response of the dubbing room should conform to the top diagram of Fig. 8.1 (reproduced from the aforementioned paper) when dubbing is done from optical prints, while its acoustic response should conform to diagram B of Fig. 8.1 when dubbing is done from magnetic release prints. Since there is a difference in the two acoustic responses, the dubbing mixer must have adjustments in his console to achieve the desired acoustic responses.

The acoustic response measurements in the dubbing chamber are to be made with one-third octave bands of pink noise, and a minimum of 5 test positions should be employed in the reverberant field of the room, and averaged.

To achieve as closely as possible a correlation between the dubbing room and the theater sound, the former enclosure should be equipped with the type loudspeakers generally used in the motion picture theater. The prevailing loudspeakers in new theaters are those built by the Altec Lansing Corp.,* called "The Voice of the Theatre."

* 1515 South Manchester Ave., Anaheim, CA, 92803

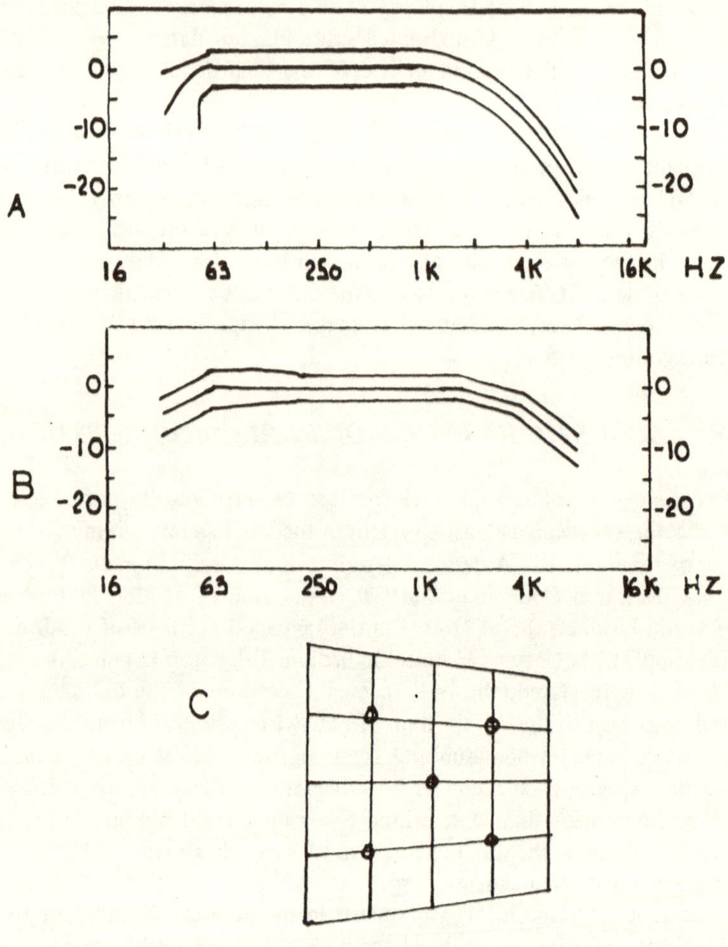

Fig. 8.1A Proposed optical/acoustic response for public and studio theaters

Fig. 8.1B Dubbing room "B"-chain acoustic response

Fig. 8.1C Recommended microphone positions for test

In control rooms in which such large theater speakers cannot be installed well, Altec 604E's or *JBL* (James B. Lansing) units in the 4330 series are generally employed.

To understand the electroacoustic requirements for a dubbing chamber, it is necessary to consider that practically all rerecording is done in this enclosure not with optical but with magnetic sound tracks. When the transfer from magnetic to optical tracks takes place, frequency-response equalization is applied to the negative of the optical tracks to compensate for slit losses, film developer characteristics, etc. Hence the magnetic/acoustic response of the dubbing room cannot be the same as the optical/acoustic response of the theater. Note that when in the theater the sound tracks are magnetic, the theater projection equipment is provided with a switching system to alter the electric response of the reproduce chain accordingly.

When dubbing is done from magnetic work prints (which later will be transferred to optical release prints) the rerecording mixer lowers the high-frequency response of his monitor to hear the sound that he expects to hear in the theater.

To measure the acoustic response of the dubbing room, no test film is required, the pink-noise generator being fed into the audio bus of the system. Preferably this signal is fed in octave bands into the system. When one-third octave bands are used, more than five microphone posts should be employed in the room because of the greater sound pressure level variations resulting from the narrower bands. This is required also when a wide-band pink noise is fed into the system and the acoustic response in the room is measured with one-third octave band filters.

The SMPTE 35 MM photographic multi-frequency test film carries the code designation P35-MF. A NFTU (Nordisk Film/TV Union) pink noise film OA 351 is available from The Danish Filmstudio.*

* 52 Blomstervaenget, DK-2800 Lyngby, Denmark.

Chapter 9

Plumbing Noise Control

No other noise in a completed building is quite as difficult to correct as the swishing and rippling of water in pipes when they were not properly installed during building. One can place seals on jambs when the doors are loose and one can even add another window pane to reduce the transmission of noises into a room, but to eliminate water noise requires the ripping out of finished wall and floor surfaces to get to the pipes which must be isolated from the framing structure so that they cannot set the large building panels into vibrations tied to the structure.

Below are the most important control measures to prevent water noise from becoming audible in a building:

1. Mount all pipes compliantly to studs and joists. Fastening fluid conduits directly to supporting surfaces is probably the most frequent cause of audible water noise in a building. While the noise radiated directly from a pipe may be practically inaudible because of the small radiating surface represented by the conduit and its relatively small wall movements, the moment this source makes solid contact with a large "diaphragm," this surface can produce large air volume displacements and therefore considerable audible noise. This investigator, on his visit to an architect or builder, often demonstrates this effect by striking a small tuning fork against the knee, which produces only a very faint signal, and then placing the bottom or handle of the fork against a desk top or the floor of the room, with the result that suddenly a loud and clear ring is radiated from the surface against which the fork is pressed.

There is no reason to manufacture costly metal straps and felt or plastic isolating strips when prefabricated pipe isolators are available on the open market. Most building supply houses stock Potter-Roemer or Mason Industries pipe isolators, often in a wide variety of sizes and Neoprene thicknesses to fit pipes from ½ in. to 10 in. (2.5 to 25 cm) in diameter.

When pipes have to penetrate concrete floors they should be isolated from them by a nonhardening plastic seal. Large pipes penetrating such floors and requiring a strong support should be fastened to a clamp which rests on springs (see Fig. 9.1).

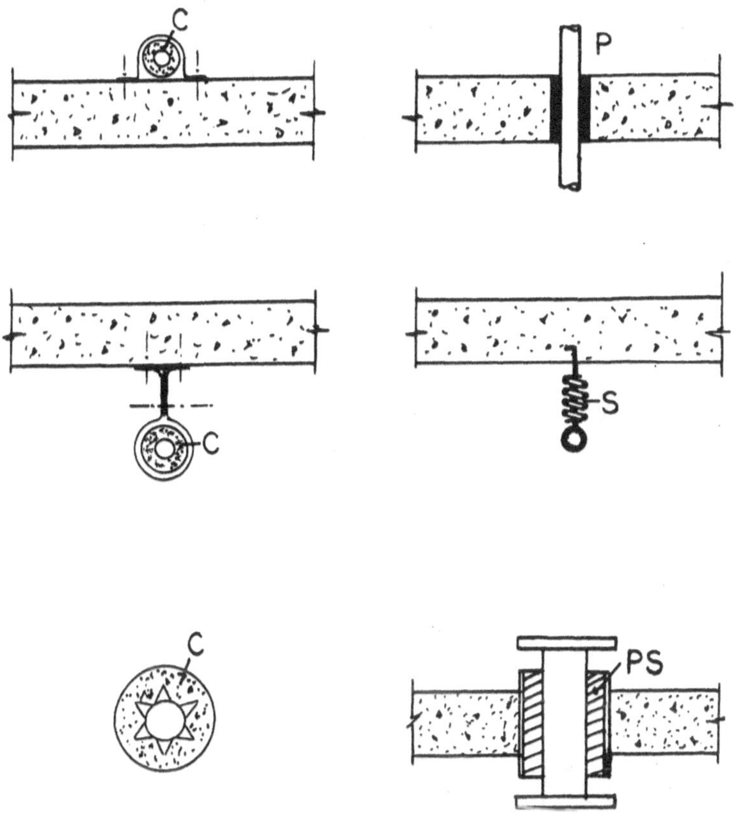

Fig. 9.1 Pipe isolators. C: cork, felt, or fiberglass; S: spring; PS: prefabricated sleeve.

2. Employ a water pressure regulator and pipes of generous size to prevent turbulent water flow.

3. Use prefabricated water-hammer arresters, but not air-filled "stubs" or short vertical pipes in the hope that the air pocket inside the pipe will act as a pneumatic spring. Such "home-built" arresters are soon deprived

of the air pocket so that they are useless as water flow dampeners.

4. Set bathtubs, toilets, etc., on a floating concrete floor instead of tying them directly to floor joists or concrete structural slabs. Use toilets with an integral tank; a so-called centrifugal water closet, instead of a stool whose water tank has to be mounted on a wall, as shown in Fig. 9.2.

5. For highly vibromotive noise sources, like washing machines, garbage disposals, dishwashers, etc., employ flexible pipe connectors for coupling supply and drain pipes to the units.

6. For pipe sizes larger than 6 in., reduction of the radiated noise may be had by using fiberglass pipe "lagging."

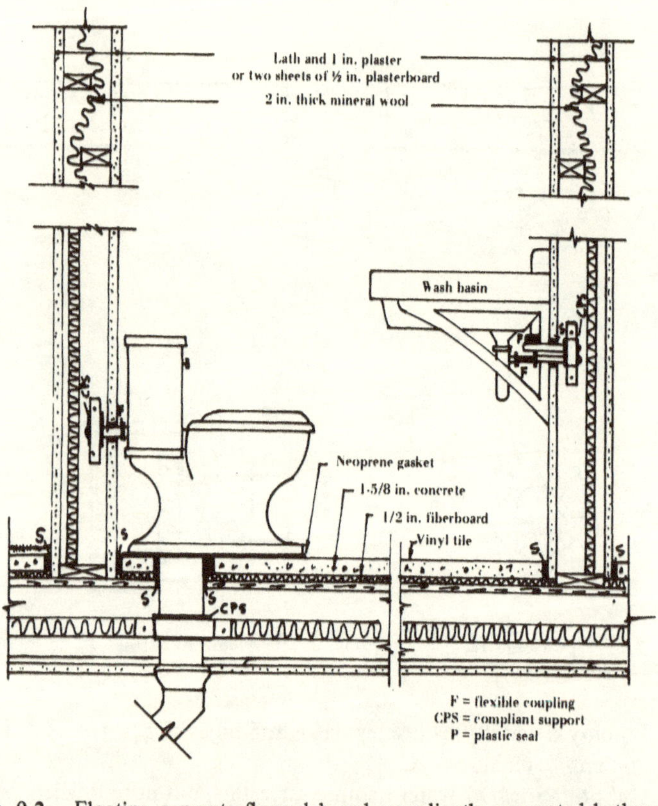

Fig. 9.2 Floating concrete floor slab and compliantly supported bathroom fixtures with resiliently mounted supply and drain pipes for the prevention of solid-born sound transmission in new buildings.

Chapter 10

Vibration Isolation

For noise-sensitive rooms, like recording studios, TV and cinema stages, the preferred location of the air-conditioning system machinery is at a distance from the building, outside, or else in the basement.

There are several reasons why the roof of a structure is not a desirable position for such an assemblance of noise sources:

1. The required sound level difference between the external noise and the desired internal one is simply too great to allow the economic construction of a roof with sufficient surface density. Thus, when the noise level near the equipment, L_0, is in the order of 90 dB in the 125-Hz octave, and the required internal sound level, L_i, at the same frequency is 38 dB, as it would have to be for a *PNC*-20, which is comparable to an A-weighted sound level of 30 dB-A, then the sound transmission loss *TL* of the roof would have to be 90 – 38 = 52 dB at 125 Hz. It is out of the question to construct such a roof since, as a monolithic barrier, it would have to weigh 700 lb/ft² (3416 kg/m²) according to $TL_{125} = 11 + 14.5 \log M = 52$, when $M = 700$ lb/ft².

Even if L_0 were lowered to 80 dB and L_i were allowed to rise to *PNC*-25 (35 dB-A), that is, 42 dB at 125 Hz, the surface density of the roof for this 38 dB sound level difference would still have to be 75 lb/ft² (366 kg/m²) comparable to 6 in. (15 cm) of concrete.

The walls of the studio to carry such a high roof load would have to be similarly high load-bearing, as would associated trusses. Note that in tilt-up construction the walls are considered nonloadbearing curtain walls.

2. Consider now the special requirements for an air-conditioning system for a noise-sensitive room:

a) To prevent turbulency at the diffusers as well as to keep the aerodynamic noise in the ducts to a minimum, the air speed should be no more than 500 ft (152 m)/min. Say 5-min air changes are required for a TV studio of 37,000 ft³ (1047 m³), then the cross-sectional area of the ducts has to be (37,000/5)(500) = 14.8 ft^2 (1.37 m^2). When there are 3 diffusers, each duct would have to be nearly 5 ft^2 (0.46 m^2) in cross-sectional area. When such large ducts are located on the roof of the studio, they become fair game for the sound transmission of aircraft noise unless the ducts are housed in. This, of course, adds further weight to the roof and makes maintenance work more difficult.

b) Because of the low air speed, fans with a low rpm are required. These have a low "driving" frequency f. For instance, if the rotary speed of a fan is 300 rpm, the driving frequency is 300/60 = 5 Hz. For any resiliently supported machinery to achieve a 15-dB vibration reduction, the resonant frequency f_0 of the mechanical system should be in the order of 1.25 Hz, so that $f/f_0 = 4$. This, in turn, necessitates a 2.6-in. static deflection on the part of the springs, or a total spring height of nearly a foot for a 20 percent compression of the springs.

It may be noted here that, in some instances, where economy was all-important, the air-conditioning system was made to operate at two speeds—a slow one during "takes," when the number of air changes was drastically reduced, and a fast one during rehearsals or during the changing of sets, when the noise levels in the stage were not overly important.

3. In many cases a 15-dB vibration reduction by resilient supports is not enough, and an inertia block of four-times the weight of the air-conditioning machinery becomes necessary to obtain another 14-dB vibration reduction according to

$$\text{Delta dB} = 20 \log \left(1 + \frac{M_b}{M_m}\right)$$

$$= 20 \log \left(1 + \frac{4}{1}\right)$$

$$= 14 \text{ dB}$$

where

M_b = mass of inertia block
M_m = mass of machinery

Fig. 10.1 shows the required static deflection in inches for various forcing frequencies to achieve the desired decibel vibration isolation shown as a parameter in the figure. The pertinent equations for the sound attenuation are

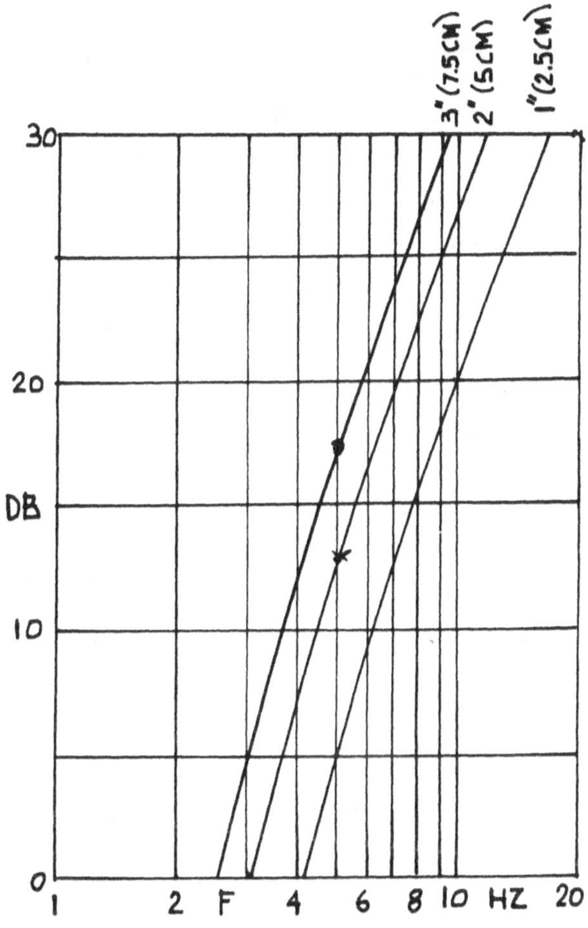

Fig. 10.1 Variation of vibration level reduction with forcing frequency F for various static deflections of the spring isolator

$$A = 20 \log \left(\frac{f^2 d}{9.8} - 1 \right)$$

$$= 20 \log \left(\frac{f^2 d'_{\prime}}{24.9} - 1 \right)$$

where

d = static deflection isolator (in.)
d' = static deflection of isolator (cm)

As an example, consider a fan with 300 rpm, so that its chief forcing frequency is 300/60 = 5 Hz. To secure a 13-dB isolation for the vibration by means of a spring isolator, a static deflection of 2 in. (5 cm) is necessary (see cross in figure), while for a 17 dB isolation a static deflection of 3 in. (7.5 cm) would be necessary (see dot in figure).

The equation for the amount of vibration isolation is

$$\text{Attenuation} = 20 \log \left(\frac{f^2 d}{9.8} - 1 \right) \text{ dB}$$

The following table gives the amount of vibration reduction due to an inertia block

M_b/M_m	A = level reduction
1	6
2.16	10
4.16	15
9	20
30	30
100	40

The pertinent equation for the above reduction is

$$A = 20 \log \left(1 + \frac{M_b}{M_m} \right)$$

The resonance frequency of the floating concrete floor due to the stiffness of the trapped air below the floor is

$$f_a = \frac{170}{(mh)^{1/2}} = \frac{60}{(m'h')^{1/2}}$$

where

m = surface density of floor (lb/ft^2)
m' = surface density of floor (kg/m^2)

h = space between floor and structural slab (in.)
h' = space between floor and structural slab (m)

For a concrete floor with a specific mass of 100 lb/ft^2 (488 kg/m^2) and a 3 in. (7.5 cm) deep air space, f_a comes to 9.83 Hz.

The effective resonance frequency of a floating concrete floor due the mechanical system characterized by its static deflection and due to the trapped air below the floor is

$$f_e = \left(f_0^2 + f_a^2\right)^{1/2}$$

Thus, when $f_0 = 10$ Hz and f_a has the same value, the effective resonance frequency of such a floor comes to 14 Hz.

Placing the air-conditioning machinery in the basement of a building is a great help in preventing the transmission of air-born as well as solid-born noise to the studios. It still requires the placement of the high noise sources onto a floating concrete floor, which may act as an inertia block, and still necessitates the resilient support of the machinery on springs with a large static deflection. Also, the walls and roof of the mechanical equipment room should employ concrete, and the interior of the enclosure should be treated with a material of high absorptivity. The distances involved between the mechanical room and the studios should be sufficiently large so that considerable noise attenuation occurs in the structures between the two enclosures.

The discussed vibration isolator static deflections pertain also to machinery spring-suspended from overhead rafters or joists, as in the case of ventilating equipment occasionally used in TV and cinema stages to draw out the fumes generated in scenes requiring smoke or fog as background. The vibrations of this equipment should not be allowed to be carried into control rooms, even though the stage is not in use during the air-exhaust period. Similar isolation measures are required when the air-conditioning system ducts are suspended from the roof of a studio, because these ducts may also be vibrating.

When the suspended ducts are relatively large, and present a parallel hard surface to the hard floor, with the consequence of flutter echoes between them for impulsive signals produced on the floor, the underside of the ducts should be treated sound-absorbently.

When vibratory equipment is mounted on a homogeneous floor or a roof below which lies a noise-sensitive enclosure, it is possible to estimate the sound pressure level in the lower room from a knowledge of the acceleration amplitude of the floor or roof when the microphone is close to the vibrating surface. At some distance from the ceiling, the sound pressure level is to an extent determined by the area of the vibrator, the total sound absorption in the lower enclosure, and the radiation efficiency of the vibrating material of which the upper floor or roof is composed.

Fig. 10.2 shows the variation of the acceleration with frequency for various sound pressure levels, *SPL*'s, developed close to the vibrating surface. Also shown in the figure are the accelerations for *PNC*-20 and *PNC*-25 (Preferred Noise Criterions 20 and 25), which correspond respectively to A-weighted sound levels of 30 and 35 dB-A. It is seen that for such noise criteria levels the acceleration of the upper floor or roof should be kept below 1 cm/s^2, which corresponds to an acceleration level of –60 dB referred to 1 g (980 cm/s^2). The equation for this condition is

$$SPL = 20 \log a - 20 \log f + 90$$

where

a = acceleration in cm/s^2
f = frequency

At some distance from the vibrator, the equation is

$$SPL' = SPL + 10 \log (4Ss/A)$$

where

S = area of vibrating surface (ft^2 or m^2)
A = sound absorption in lower room (sabins or metric sabins)
s = radiation efficiency (generally in the order of unity)

It is seen that when A is larger than $4S$, the *SPL* in the lower portions of the enclosure is less than near the vibrating surface (assuming $s = 1$). This is to be expected for the sound level at some distance from a vibrating wall or partition, because of the distance effect. However, for a sound stage with a vibrating roof or ceiling, the sound level reduction may not be as much as would be expected from similar situations in a room with a vibrating partition. The reason is that the roof of a sound stage is facing a hard, generally concrete, floor, on which the vertical sound "rays" from overhead impinge at right angles and thus are able to return vertically with little amplitude reduction. On the other hand, a sound stage or TV studio usually also carries sound-absorbent treatment on the roof or ceiling, so that some sound attenuation occurs when the sound waves have to pass first through the acoustic treatment before they can enter the room proper.

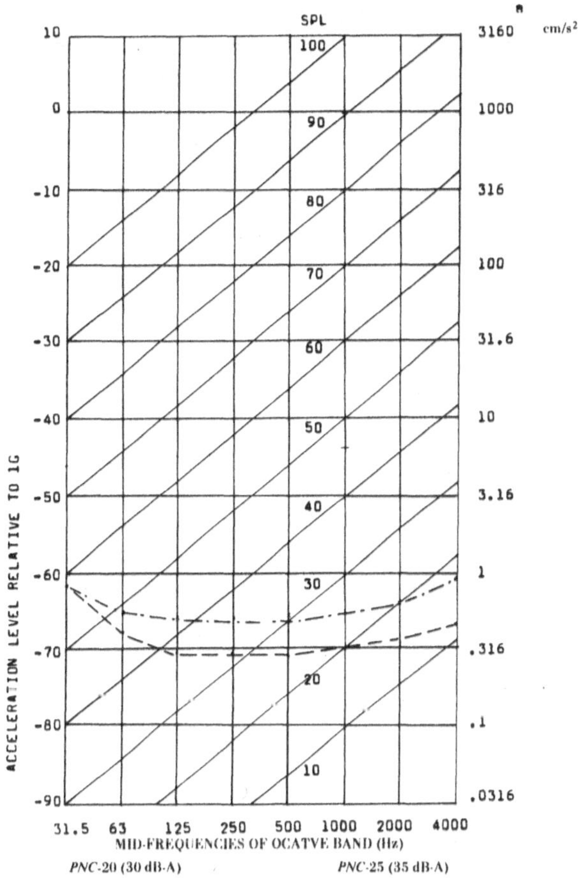

Fig. 10.2 Variation of acceleration of vibrating surface with frequency for various sound pressure levels near the surface

Because no acoustic treatment is applied to the floor of a studio, the acceleration levels for this surface should be kept within the limits discussed above.

Low acceleration levels at a prospective studio site are particularly recommended when nearby rail traffic exists. The solid-born and solid-borne vibrations from trains can travel for miles with little reduction in their amplitude when the ground is hard. The digging of ditches to prevent the transmission of these vibrations is generally of little value, unless the ditch extends completely around the studio. In a severe case it may become necessary to float the entire building on a resilient bed of Neoprene, with flexible conduits for all water and gas pipes to the structure.

The reduction of vibrations from a mechanical equipment room into a TV studio floor is generally accomplished by floating the concrete floor of the equipment room on 3 in × 3 in. × 3 in. fiberglass pads and by casting the studio concrete floor on a 6 in. thick layer of sand or decomposed granite and not allowing the floor to contact the foundations of the building. Almost all high-quality TV cameras now employ three image-forming Plumbicon or Saticon tubes. Behind the single lens of the electronic camera is a beam splitter that trisects the image into three paths. In each path is a color filter, red, green, or blue, before the light is sent to the target screen of the tube. It becomes important for each of the three tubes that the color components coincide into a single image, a process called registration. To obtain precise registration, the tubes and prism assembly must be ruggedly constructed and shock-mounted, so that vibrations into the camera cannot cause color fringes. This objectionable vibration effect is, of course, further reduced by minimizing the transmission of such mechanical disturbances into the floor of the studio.

In the following a set of calculations is presented for the determination of the type, number, and placement of 3 in. × 3 in. × 3 in. fiberglass vibration isolators, Neoprene-coated, and preferably precompressed. The same type of calculations can be used for steel springs, cork pads, and other isolators intended to be used elastically to support the concrete floor of the mechanical equipment room.

Fig. 10.3 shows two different details of construction for a floating concrete floor.

Fig. 10.4 shows static deflection versus load curves for 1 in. thick Peabody Noise Control Co. isolation pads. For a specific load of, e.g., 31.2 lb/in.^{2} (2.2 kg/cm^2) on a type L pad, the deflection on a 1 in. (2.54 cm) thick isolator is 0.22 in. (0.56 cm). The corresponding static resonance frequency for this mechanical system is

$$f_0 = \frac{3.13}{d^{1/2}}$$

$$= \frac{3.13}{0.22^{1/2}}$$

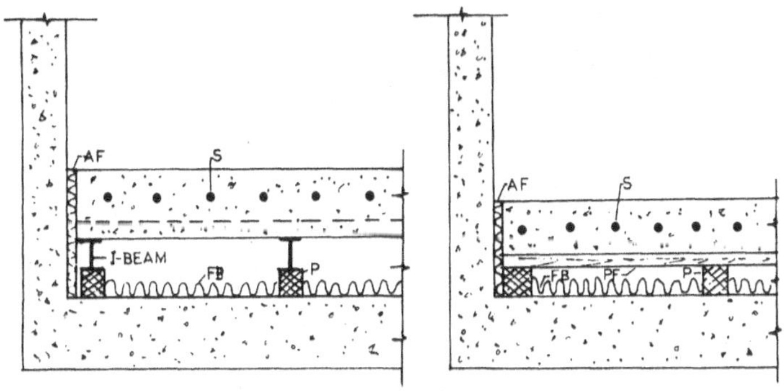

Fig. 10.3 Floating concrete floor on I-beams and metal decking (left) and on plywood formboard; FB: 2 in. thick fiberglass board; P: 3 in. × 3 in. × 3 in. fiberglass isolation pads; S: steel reinforcing rods; AF: asphalt-impregnated fiberglass board; PF: exterior plywood formboard.

$$= 6.67 \text{ Hz}$$

In the field, where dynamic conditions exist, the resonance frequency could be twice as high, or 13 Hz. Hence it is desirable to introduce an "ignorance factor," as is done in structural engineering. This may be had by employing a 2 in. thick pad, or even a 3 in. thick one.

Similar considerations hold for so-called housekeeping pads on which machinery is placed in a mechanical equipment room. Because of the extra mass of the 6 in. or 10 in. (15 cm or 25 cm) high concrete pad and the heavy machinery on the pad, it becomes necessary to calculate the amount of additional elastic isolators under the pads to prevent excessive shearing forces from being applied to the floating concrete floor. This often becomes a problem for the structural engineer to consider, and he should work closely with the acoustic consultant on the project in this matter.

Another problem which often arises when an enclosure is to be erected on a floating concrete floor is the lateral support of the studs or the concrete block wall. For the reduction of earthquake effects a wall isolator is called for in such cases, that is, an elastic horizontal brace between the external structural wall of the building and the internal wall of the studio. Fig. 10.5 shows one form of such a compliant "strut" or support. These wall isolators should not be placed haphazardly along the walls, but in strict compliance with their compressive and tensile strengths, as determined by the manufacturer. Also, an additional sound insulation measure for such

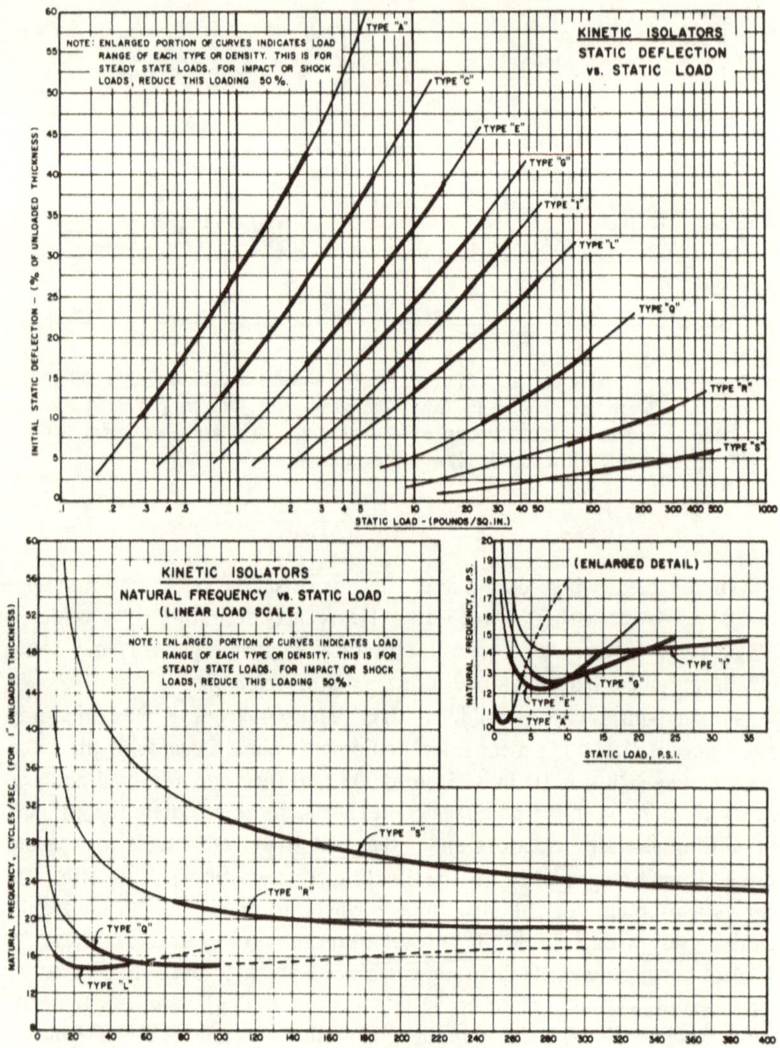

Fig. 10.4 Elasticity data for precompressed fiberglass isolators

a construction consists in the introduction of a generous amount of sound-absorbent material in the interspace to prevent the undesirable sound-transmissive effects of space resonances or standing waves.

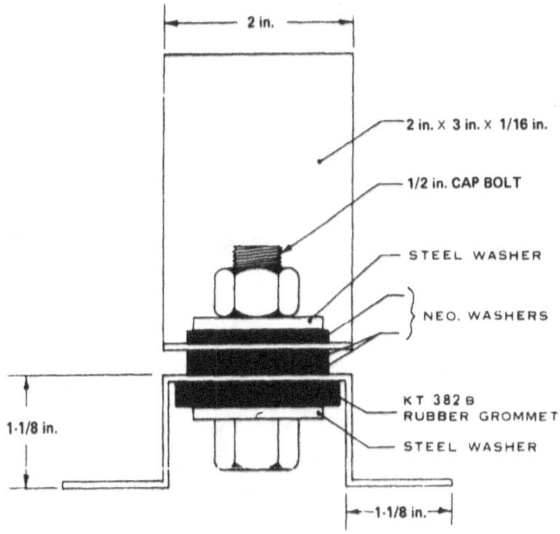

Fig. 10.5 Resilient wall mounting

Chapter 11

Suspended Acoustic Ceiling

A suspended inverted T-grid system with sound-absorbent lay-in panels represents an "integrated ceiling" in modern architectural practice. It is popular not only because the 12 in. to 36 in. airspace between this ceiling and the overhead structural slab can be used to hide pipes, conduits, recessed lighting fixtures, and ducts, but also because the plenum can be employed as either an air intake or air exhaust passage, thus saving a set of air-conditioning system ducts.

However, unless a divider with a duct silencer or a divider with sound traps is built above the partition between two adjoining rooms, cross-talk can take place across the space above the partition. Fig. 11.1 shows a divider with a duct silencer, which is essentially a short duct with built-in sound traps, which allows air passage but attenuates acoustic signals as they move along the convoluted and sound-absorbent channels. Fig. 11.2 represents a divider-trap combination which acts much in the same way as the duct silencer built into a solid divider.

When it is desired, as may be in schools, to change the size of a classroom by moving a demountable partition, such sound traps are less effective because the partition may come to lie at the "wrong" side of the divider. In such a case, sound-retardant elbows may be installed directly above the air-intake or air-exhaust opening in the ceiling, as shown in Fig. 11.3.

Besides the advantages already listed for a suspended sound-absorbent ceiling, there is also the increased absorption produced by the acoustic lay-in panels with their deep airspace between themselves and the overhead structural slab, particularly at the low frequencies. However, this construction also generates sound leakage into the plenum, which reduces the sound transmission between two adjoining enclosures, even when the plenum is not employed for an air exhaust duct. Hence, when the partition between the adjoining rooms is of heavy construction, with, e.g., a

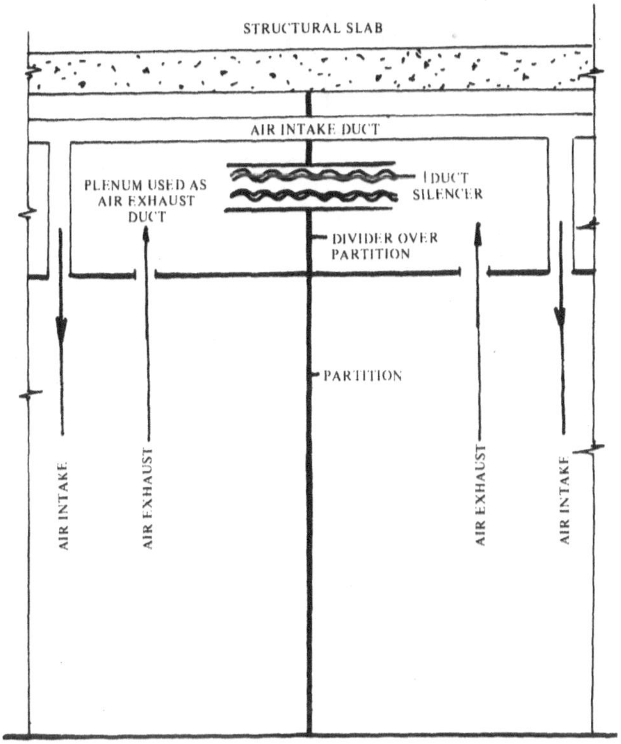

Fig. 11.1 Duct silencer in divider above the partition between two adjoining rooms prevents cross-talk between them

sound transmission loss *TL* of 50 dB at 500 Hz, this insulation is made much weaker by the ½ in. (1.25 cm) thick acoustic lay-in panels usually employed for such suspended ceilings.

There are essentially three means available to increase the *TL* across the ceiling. The first was discussed previously, and consists of employing a divider above the partition with or without a duct silencer in the divider. The second means lies in laying a 2 in. (5 cm) thick fiberglass board on top of the lay-in panels to seal the leaks between the panels and the inverted T-grid system, and sometimes also to seal the leaks along the acoustic panels when no splines or kerfs are used at the panel joints. The third method of increasing the *TL* of the ceiling is to top the acoustic lay-in

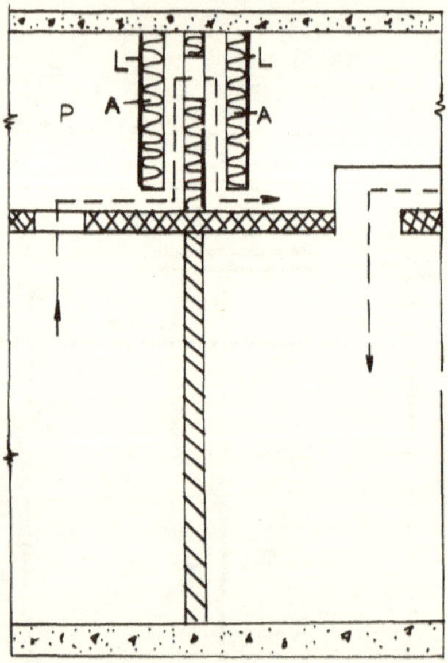

Fig. 11.2 Sound trap in plenum when it is used for an air intake or exhaust channel. L: sheet lead or plasterboard; A: absorbent; P: plenum

panels with ½ in. thick plasterboard. This improves the *TL* of the relatively soft acoustic panels by increasing their effective mass. This, in turn increases their pressure against the T-channels where the two touch, thus avoiding sound leakage between them when either the channels or the panels are warped.

Fig. 11.4 shows the effective 500-Hz *TL* for various ceilings.

It is seen that a divider above the partition, with either a duct-silencer in the divider or a home-built sound trap in it, is superior to sound-attenuative elbows and no divider. The sound-attenuative elbows merely close the ceiling openings shown in Diagram 1. and thus make the *TL* of the ceiling equal to that without an opening, illustrated in Diagram 2. The divider, on the other hand, stops the passage of sound across the partition, except for the little which can be transmitted through the sound traps in the divider.

A similar situation prevails in the case where the acoustic lay-in panels are topped by ½ in. (1.25 cm) plasterboard or plywood. The extra ceiling mass and reduced sound transfer through the joints provide 10 dB greater

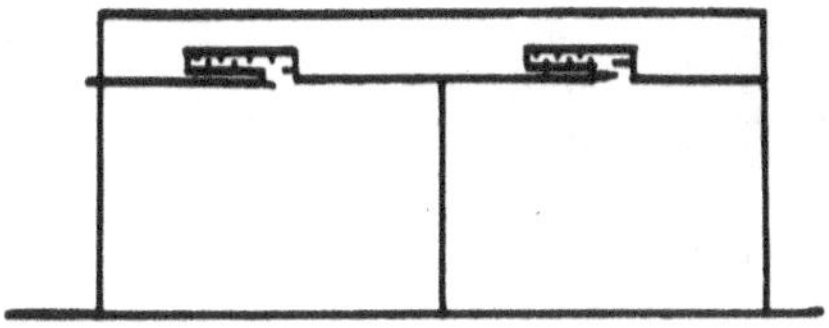

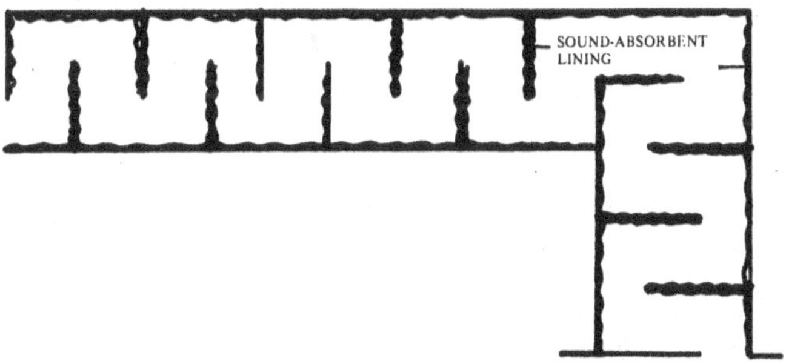

Fig. 11.3 Baffled and absorbently treated air-conditioning system duct elbows to be installed in ceiling above which the plenum is used as either an air-intake or exhaust duct

sound attenuation when no ceiling openings exist, as seen by comparing Diagrams 2 and 6. The total *TL* of the suspended ceiling shown in Diagram 6 can be made still greater by introducing sound-absorbent material in the plenum, which is "live" when plasterboard faces the plenum (unlike in the case of the ½ in. acoustic lay-in panels, which tend to provide a "dead" plenum).

Even when the plenum above the ceiling is not used for an air exhaust duct, cross-talk may be experienced between adjacent rooms when the same duct is used to feed air or exhaust air from the rooms, unless a duct silencer is introduced in the duct between the two enclosures.

DIAGRAM	CONSTRUCTION	DESCRIPTION	TL WALL	TL CEILING	TOTAL TL
1	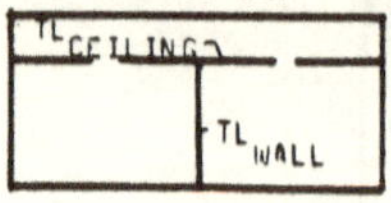	½ IN. ACOUSTIC LAY-IN PANELS	50	15	17
2	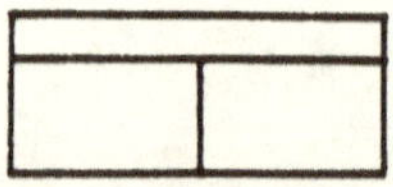	SAME AS DIAGRAM 1 BUT NO HOLES IN CEILING	50	40	42
3	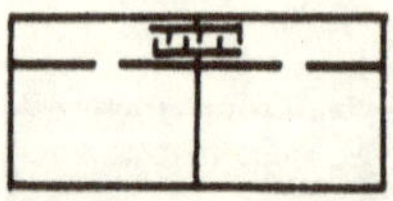	SAME AS DIAGRAM 1 BUT WITH 5-FT LONG DUCT-SILENCER IN DIVIDER ABOVE PARTITION	50	50	50
4	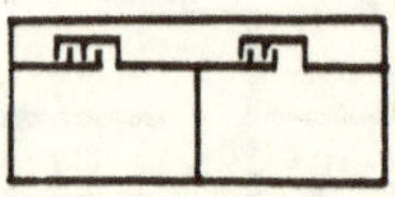	SAME AS DIAGRAM 1 BUT WITH SOUND-ATTENUATIVE ELBOWS IN CEILINGS	50	40	42
5	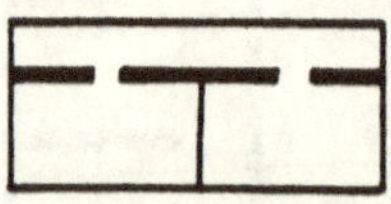	½ IN. ACOUSTIC LAY-IN PANELS TOPPED BY ½ IN. PLASTER-BOARD	50	16	18
6	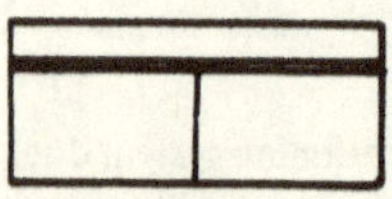	SAME AS DIAGRAM 5 BUT NO HOLES IN CEILINGS	50	50	50
7	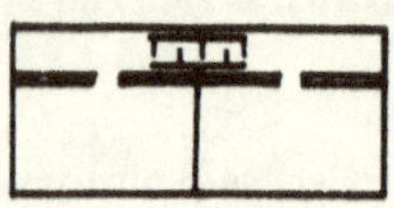	SAME AS DIAGRAM 5 BUT WITH 5-FT LONG DUCT SILENCER IN DIVIDER ABOVE PARTITION	50	53	51
8	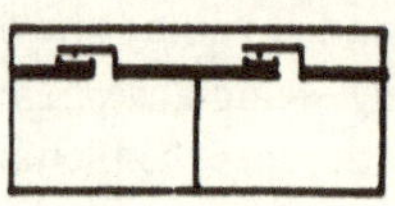	SAME AS DIAGRAM 5 BUT WITH SOUND-ATTENUATIVE ELBOWS IN CEILINGS	50	50	50

Fig. 11.4 500-Hz sound transmission loss between adjoining rooms with suspended acoustic ceilings with plenum used as air-exhaust duct

Chapter 12

Elevators

For raising or lowering people or freight in a building, elevators are employed for heights in excess of some 10 or 15 ft, while escalators may be used for lesser elevation differences. There are essentially two types of "lifts:" electrically and hydraulically operated units. The former are employed for building heights in excess of 50 ft, and the latter for lower levels.

In a building with noise-sensitive rooms, the elevator shaft and the motor room should not adjoin such enclosures, because of the structure-borne sound which may be transmitted into them by the cab and the machinery. When space limitations exist, the motor and its associated winch should be placed on a resiliently supported inertia block, and when the respective building ordinance allows it, the guide rails in the shaft should also be vibration isolated.

Hydraulic elevators are by far the most quiet, because the plunger in its cylinder is below ground at a depth approximately equal to the height that the elevator travels. The ram with its cab is raised by forcing oil into the thick-walled steel tubing known as the pressure cylinder, and is lowered by releasing the oil from its container. There are no wire ropes, overhead equipment, penthouse, counterweights, etc., so that vibration effects are limited to the guide rails in the shaft.

Noise effects by elevators may generally also be reduced by restricting their vertical speeds to less than 200 ft/min, by installing door bumpers, and by providing properly sized ventilation openings in the shaft to avoid the hissing sound often heard when an elevator door opens.

Chapter 13

Interior Decor of Studios

It is the prerogative of the architect and interior decorator to choose the finish of the walls, ceiling, and floor of a studio, possibly with the assistance of the acoustic consultant of the project when sound-absorptive or sound-reflective areas are desirable at given locations of the enclosure. Often such materials are selected from catalogs or folders which contain only a meager description of absorptive and flame-resistance qualities of a fabric. What can be the absorptivity of a cloth 0.010 in. or 0.020 in. thick? It would have to be folded many times in the form of a curtain to absorb effectively—and who knows its absorptivity then without a test when hung a given distance from a hard wall?

In respect to the fire-resistance, an equal lack of information is often provided by the manufacturer of the product. A catalog or folder may describe this quality as "Class II, with a flame-spread index of 30, fulfilling Federal Specification SS-S-118 as Class 75." Anyone who has ever investigated ASTM E-84 describing the "tunnel test method" for flame spread must wonder what is meant by the rate at which flame will travel across an exposed surface. This value compares the rate of flame travel on a selected material within the limits of untreated lumber, rated 100, and noncombustible cement-asbestos board classified as 0. To come to graphic grips with the problem, this investigator has at numerous times held a small piece of the selected fabric on the tip of a pocket knife and held a match under it. Some very surprising results can in this manner be obtained regarding the combustibility or flame-support of a chosen material, whether it is a wool fabric or a plastic one. In 1972 an extensive fire on a set containing polyurethane material ruined an entire sound stage at the United Artists Studios in Hollywood when a lighting fixture was accidentally turned over and set the polyurethane on fire. The knife tip test would have immediately ruled this material out for a picture set component.

The situation is more significant when costs of completely incombustible sheets like glass cloth are compared with linen or plastic materials. Colorful and thick glass cloth, like prestretched "Soft Flex" weighing as much as 10 oz/yd^2, is no more expensive than velvet or tapestry materials.

There are also many loudspeaker grill cloths which stand up under the knife-tip test which are highly decorative and sound-transparent. But even these products, when they contain paper threads, may burn readily, so that they should always be tested in the manner described above.

Interior decorators often select the carpets intended for the covering of the floors in control rooms and studios, which is as it should be. But there are literally thousands of different types of carpets and pads or underlays on the market, few of which are rated in respect to their absorptivity at different frequencies. In such instances, the acoustic consultant should try to obtain samples of the product and make an absorptivity test with an impedance tube, as described in ASTM Standard C-384-58. Such a test may even be performed on a carpet installed in a studio. This does not mean that, if the covering is found wanting, it has to be replaced. Instead, additional absorptive material may be installed on the walls or ceiling of the room.

Chapter 14

Sound Power Versus Sound Pressure

Occasionally the terms sound power level and sound pressure level are not clearly understood or sharply differentiated in conversation. In noise control work and the acoustic design of rooms it is necessary, however, to make a definite distinction between the two quantities, particularly since they complement each other and together provide vital information about an enclosure as well as about a specific source of sound.

Sound power is, simply, the total sound energy radiated per second by a source. The unit is the watt, equivalent to 10^7 ergs/s. The direct measurement of sound power is complicated because it involves the assessment of the time integral of the product of the sound pressure, a scalar, and particle velocity, a vector. In the laboratory this is usually accomplished by a small pressure microphone next to a velocity microphone, each with its own amplifier, a phase correcting network to compensate for the phase relationships in the microphones and the amplifiers, and an electric power indicator (often a thermocouple).

Since the sound intensity is proportional to the mean-square sound pressure, it is more convenient to measure this quantity in the open about a hemisphere. As an example, assume one has an outdoor hemisphere *S* of 100 m^2 (4 m radius), which is to be referred to a reference surface of 1 m^2, and that the sound source whose power level is to be determined radiates energy more or less uniformly about this hemisphere and is located on the ground. One measures the sound pressure level *SL* in the center of four equal areas of 25 m^2. The *SL* of each area is then added to 10 log 25 = 14 dB. Thus, if SL_1 = 100 dB, the sum comes to 114 dBp (where p refers to 1 pW, or 10^{-12} W). The 4 power levels *WL* are then added logarithmically to obtain the total power level of the source. This measurement procedure may be carried out at different octave or one-third octave levels to secure the sound power characteristic of the source, or one may measure the A-weighted *SL*'s at the four different microphone positions and thus achieve the A-weighted power level of the source.

One may also assess the power level of a source in the reverberant field of a room, but such measurements are often more complicated because of the spatial variation of the sound pressure level in an enclosure. Such tests require also the evaluation of either the reverberation time of the enclosed space or its total absorption, preferably at different frequencies when the sound power level characteristic is to be established. Some laboratories also employ rotating vanes or paddles in the room to "stir" up the sound for the purpose of reducing the interference effects in the reverberant field.

What are the uses of the sound power level of a source? Two examples will be given below.

1. Assume a pistol shot in the open produces a sound pressure level of 120 dB-A at 10 ft (3.1 m) from the gun. If the same gun is shot off in a room of 5000 ft^3 (141.5 m^3) volume, with a reverberation time of 1 s in the mid-range of the audio spectrum, what will be the *SPL* in the reverberant field of the enclosure?

To discover this the following equations are used:

$$(1) \quad WL = SPL + 20 \log R - 10 \log Q + 0.7$$

$$= SPL + 20 \log R' - 10 \log Q + 11$$

$$(2) \quad WL = SPL + 10 \log V - 10 \log T - 29.3$$

$$= SPL + 10 \log V' - 10 \log T' - 14$$

where

WL = sound power level referred to 10^{-12} W
SPL = sound pressure level
Q = directivity factor of source; it is 1 for a nondirectional source, 2 for a source radiating into a hemisphere, and 4 for a source radiating into a quarter sphere, etc.
T = reverberation time
R = distance from source (ft)
R' = distance from source (m)
V = volume of room (ft^3)
V' = volume of room (m^3)

From equation (1) it can be seen that *WL* is 140.7 dBp. When this value is introduced in equation (2), the *SPL* of the gun shot in the reverberant field becomes 132 dB. This indicates that a much smaller charge should be exploded in the room to prevent possible damage to the microphone.

2. Assume an air-conditioning system distribution fan was measured by its manufacturer to have a noise emission level of 8.5 bels, as measured with the A scale of the common sound level meter. The ISO Technical Committee 43 adopted this designation to prevent confusion between the similarity between sound power levels and sound pressure levels. Since 1 bel equals 10 decibels, for the A-weighted sound power level of the fan under consideration, $WL = 85$ dBp can be written. This is the total acoustic power, so that at both the intake and the exhaust of the fan the sound power level is 3 dB lower, or 82 dB-A.

When the area S of the duct which is to be connected to the fan is 10 ft^2, the SPL at the duct opening is given by

$$\begin{aligned} SPL &= WL - 10 \log S + 10.6 \quad \text{or} \quad = WL - 10 \log S' + 0.3 \\ &= 82 - 10 \log 10 + 10.6 \\ &= 82.6 \text{ dB-A} \end{aligned}$$

where S or S' = duct area (ft^2 or m^2).

It is seen that one of the advantages of expressing the noise level of a source in terms of its power level is that one does not have to note any distance from the source. A noise sound pressure level of, say, 85 dB-A is meaningless unless one also notes the distance from the source where the measurement was made.

When, in the open, a loudspeaker radiates a complex signal and a second unit is made to radiate a like signal in the same direction, so that at the point of observation twice the sound power prevails, then both the sound power level and the sound pressure level increase by 3 dB. When we have N like sources, e.g., automobiles or pumps, close enough together so that their distance to the microphone is essentially the same, then the sound pressure level of the noise as well as its ound power level increase by $10 \log N$. Thus when N is 10, these levels increase by 10 dB. This is because of the constantly changing phase relationships between all the components which make up the noise. The mathematics of the subject is contained in Article 42a in Volume 1 of Lord Rayleigh's "The Theory of Sound."

When, in the open, a loudspeaker radiates a single frequency, and a second identical emitter is placed next to the first and is made to radiate a sinusoid of the same frequency and of equal amplitude and phase to the first, then at a given central axis point of observation, where again twice the sound power prevails, the sound power level increases by 3 dB but the sound pressure level increases by 6 dB. This is not true when the second speaker radiates a frequency different from the first, even though it may radiate the same amount of power; in this case, both the sound power and the sound pressure level increase 3 dB.

The concepts of sound power and sound power level are useful also in evaluating the performance of a loudspeaker. Assume a loudspeaker with an efficiency of 10 percent. This means that when 10 W of electric power is delivered to the unit it will radiate 1 W of acoustic power, or an acoustic power level of 120 dBp, according to

$$\begin{aligned} WL &= 120 \log W + 120 \\ &= 120 \log 1 + 120 \\ &= 120 \text{ dBp} \end{aligned}$$

When such a unit is built into the wall of a stage in an outdoor theater, so that Q will be 2, and the theater is 150 ft (45.7 m) long, what will be the mid-range *SPL* at the rear seats?

$$\begin{aligned} SPL &= WL - 20 \log R + 10 \log Q - 0.7 \\ &= WL - 20 \log R' + 10 \log Q - 11 \\ &= 120 - 20 \log 150 + 10 \log 2 - 0.7 \\ &= 79 \text{ dB} \end{aligned}$$

where R' = distance (m).

If the *SPL* at this distance is to be 90 dB, *WL* must become 120 + 11 = 131 dBp. This calls for an acoustic power of

$$\begin{aligned} W &= 10^{(131-120)/10} \\ &= 12.6 \text{ W} \end{aligned}$$

Hence, at least 12 such loudspeakers will be necessary, along with an electric power of 126 W.

The air volume displacement of a baffled piston as a function of the sound power radiated by it at a given frequency f is

$$V = \frac{dD^2\pi}{4} = \frac{90{,}432W^{1/2}}{f^2} \quad (\text{in.}^3)$$

$$V' = \frac{d'D'^2\pi}{4} = \frac{90{,}432W^{1/2}}{f^2} \quad (\text{cc})$$

$$dD^2 = \frac{115{,}200W^{1/2}}{f^2} \quad (\text{in.}^3)$$

$$d'D'^2 = \frac{1{,}888{,}128W^{1/2}}{f^2} \quad (\text{cc})$$

where

d = piston displacement from mid or rest position (in.)
d' = piston displacement from mid or rest position (cm)
D = piston diameter (in.)
D' = piston diameter (cm)
W = sound power (W)
V = air volume displacement (in.^{3})
V' = air volume displacement (cc)

As an example, what is the required air volume displacement of a baffled loudspeaker at 50 Hz when the maximum excursion of the cone is to be 0.5 in. and the power output is to be 16 W?

$$V = \frac{90{,}432 \times 16^{1/2}}{50^2}$$

$$= 144 \text{ in.}^3$$

$$D = \frac{339W^{1/4}}{fd^{1/2}} \text{ in.} \qquad D' = \frac{1374W^{1/4}}{fd'^{1/2}} \text{ cm}$$

$$= \frac{339 \times 2}{50 \times 0.7}$$

$$= 19 \text{ in.} \qquad = 48 \text{ cm}$$

The required sound power may also be expressed as a function of the sound pressure level *SPL* at a distance R in feet (R' in meters) and the directivity factor Q. Thus

$$W = \frac{1.16R^2 \times 10^{SPL/10}}{Q \times 10^{12}} = \frac{12.48R'^2 \times 10^{SPL/10}}{Q \times 10^{12}}$$

$$W^{1/2} = \frac{1.07R \times 10^{SPL/20}}{Q^{1/2} \times 10^6} = \frac{3.53R' \times 10^{SPL/20}}{Q^{1/2} \times 10^6}$$

Therefore

$$V = \frac{90{,}437 \times 1.07R \times 10^{SPL/20}}{f^2 Q^{1/2} \times 10^6} \quad \text{or}$$

$$V' = \frac{90.437 \times 3.26R' \times 10^{SPL/20}}{f^2 Q^{1/2} \times 10^6}$$

For a baffled loudspeaker with $Q = 2$, radiating 16 W at 50 Hz, the *SPL* at 100 ft is 94.4 dB. The required volume displacement of the cone is

$$V = \frac{90{,}437 \times 1.07 \times 100 \times 10^{4.72}}{50^2 \times 2^{1/2} \times 10^6}$$

$$= 144 \text{ in.}^3 \text{, as before.}$$

When the Q of a reproducer can be made smaller, as for instance with a long exponential horn, the required sound power and cone excursions can be less for a specified *SPL* at a given distance.

The directivity factor of a loudspeaker radiating into a spherical segment with a vertical angle a and a horizontal angle b is

$$Q = \frac{180}{\sin^{-1}(\sin a/2 \times \sin b/2)}$$

Thus, when the horizontal angle is 90° and the vertical angle is 45°, Q becomes 11.

The directivity factor of a radiator increases with the frequency. In the case of a baffled 16-in. (40 cm) diameter piston, for instance, its Q is 3 at 100 Hz, 5.5 at 500 Hz, and 11.5 at 1000 Hz. The Q also increases with the diameter of a vibrating piston, for which reason a long exponential horn is generally chosen for high Q's at low frequencies, since the mouth of the horn may be said to represent the area of a vibrating piston.

PART II

STUDIOS

Chapter 15

Control Rooms

Control rooms for music-recording studios are for the purpose of regulating the quality of the recorded program rendered in the studio. Often, too, nowadays, they serve as the mixing locales of synthetic music sources and a wide variety of sound tracks recorded elsewhere. The operations may involve level adjustments, the addition of a reverberatory note to the renditions, tonal modifications by means of equalizers, the use of compression, limiting, or expansion of various passages, and other checks and modifications.

There are no hard-and-fast rules regarding the volume which a control room should have in relation to its associated studio. For smaller studios, up to 50,000 ft^3, a control room with a volume equal to 10 percent of that of the studio is sometimes suggested; for larger studios a figure of 5 percent and for studios of less than 5000 ft^3 a figure of 20 percent is generally found acceptable.

Nor are there rigidly binding rules in respect to the floor covering of a control room. Some organizations prefer wall-to-wall carpeting of the nonelectrostatic type to prevent shocks, while others employ Nylon flooring for the easier wheeling about of cart-mounted recorders, synthesizers, and other equipment. This element is a functional requirement, and should be decided by the studio personnel. Satisfactory rooms with either carpet or plastic flooring can be designed.

The interior decor of the enclosure should always be provided by an expert in this field, particularly in respect to the color scheme of the room.

Control room acoustics have come to be considered equal to, and by many recording engineers more important than the acoustics of the main studio in which the musical program originates. This is due to the prevalent practice of combining numerous sound tracks in such a room towards a final mix of great sales appeal. When 16, 32, and even 64 music renditions are electronically joined, the significance of any one track diminishes, and it is the union of the many signals which comes to determine the commercial success of the ultimate product. This is obviously also true with "synthetic music" generated by Moogs and other devices, where no studio is involved, and true also when tracks from various studios are to be blended into a decisive tonal work.

The following represents a summary of the acoustic requirements of such an enclosure.

1. *Shape of room.* The plan or horizontal cross section of the enclosure should not be rectangular but trapezoidal or quasi so, that is, the frontal sidewalls should not be parallel but should be slanting linearly or curvilinearly like a horn. This tends to avoid coincidental reinforcement of the normal modes in the room and will certainly prevent flutter echoes when these sidewall sections are made reflective. There is no way to eliminate normal modes of vibration in a closed space, because they are a natural phenomenon with their number dependent only on the volume of the room for any one frequency interval.

Fig. 15.1 shows the average frequency space between modes for a 5000 ft^3 control room. As an example, at 21 Hz, the modes are spaced

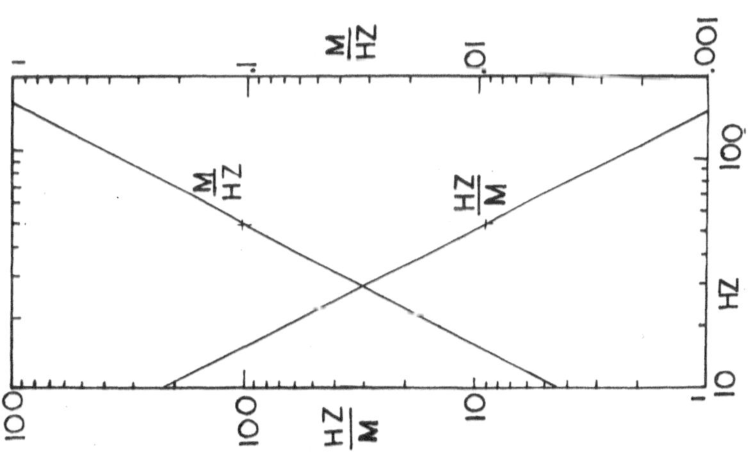

Fig. 15.1 Average frequency space in hertz between modes (M) for a control room of 5000 ft^3 volume, as a function of frequency. Ex: At 50 Hz, a frequency space of 9.1 Hz exists between modes. Conversely, only 0.11 modes exist per hertz at 50 Hz in the room.

10 Hz apart; at 120 Hz, only 1 Hz apart.

The frontal part of the horizontal cross section of the room should display some reflective panels, not unlike the area of an exponential horn near the "throat" of the unit. This tends to direct sound towards the mixer which might otherwise not be heard by him, because of the directionality of emission associated with every loudspeaker, and also constitutes more acoustic load for the low-frequency drivers of the reproducing system.

The vertical cross section of the room should similarly show a reflective area near the frontal part, between the console and the loudspeakers. Such splays, sometimes called tier drop ceiling, assist sound direction in a manner similar to that discussed above for the slanting and reflective sidewall panels. Care should be taken, when the ceiling splay or splays are flat, that their angle of orientation is not the same as that of the sloping console panel, since this would introduce an undesirable parallelism between these hard surfaces. In this respect it may also be noted that when such a parallelism exists between the flat front wall and the console, these areas should be treated sound-absorbently.

The rear sidewall portions and the rear part of the ceiling should be designated moderately absorptive which implies a degree of absorption which provides the optimal reverberation time for the enclosure after taking into account the highly absorptive rear wall (see below).

2. *Rear wall.* This area should receive an abundance of sound absorption so that a minimum of the four first-order reflections (the two sidewall, the floor, and the ceiling reflection) are returned to the recording or balance engineer. See Fig. 15.2.

3. *Low noise level.* For a control room, an A-weighted sound level of 30 dB-A is suggested, although it is not always possible to obtain it on account of the various pieces of recording equipment which have come to be located in this room. This does not mean that the noise from the outside of the building or the rumble of the air-conditioning system inside should not be kept to a very low order. This is for the reason that there are many rehearsal periods in the control room when the recording equipment is not in operation, and during these quiet periods the recording engineer sets the general trend of the mix which he or she wishes to incorporate in the product. Hence, even though the dynamic range of the signal in the control room during the final "take" may be in the order of 50 or even only 40 dB, the signal-to-noise ratio on the tape may be as much as

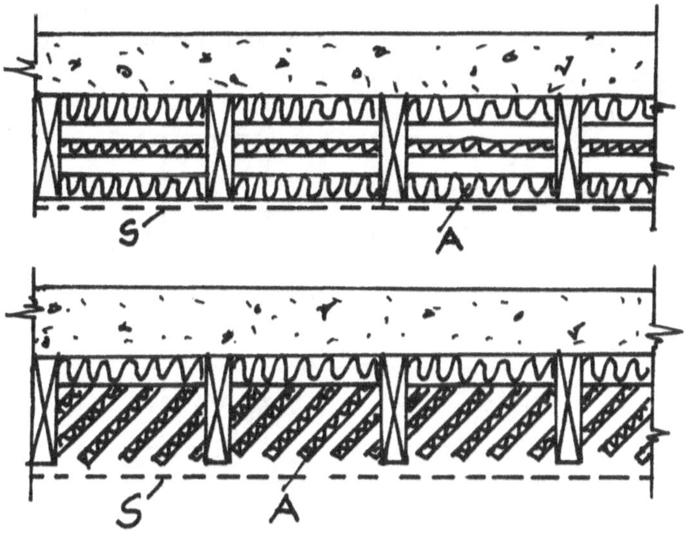

Fig. 15.2 Anechoic treatments for the rear wall of control room. S is decorative screen, like loudspeaker grill cloth, to protect 2 in. (5 cm) thick sound-absorbent fiberglass boards marked A.

60 dB.*

Nothing is gained by specifying the noise level limit in a room using a NC (Noise Criterion) curve. According to this method, a noise is not acceptable when any part of the spectrum exceeds this limiting curve, no matter how narrow the frequency band which surpasses it is. But then, a noise is also unacceptable with a spectrum equal to that of the pertinent *NC* curve but which slightly transcends this curve. Yet the two noises, carrying the same rating number, may differ widely in their A-weighted sound levels, as shown in Fig. 15.3.

4. *Frequency response.* The variation of the reverberation time with frequency should be flat between the lowest frequency which is desired for recording and, e.g., 4000 Hz. Beyond the latter frequency it is very difficult to maintain the reverberation at a constant value, because of the

* We assume here that some sort of noise reduction is employed in the multiple track recording. The reason for this is that when, e.g., 20 tracks are combined, each with a signal-to-noise ratio of 60 dB, the presence of the 20 noise "floors" lowers this ratio by 10 log 20 = 13 dB.

inevitable air absorption in the enclosure. A relative humidity of 50 percent is recommended for the room, because a lesser humidity increases this molecular conversion of sound energy into heat.

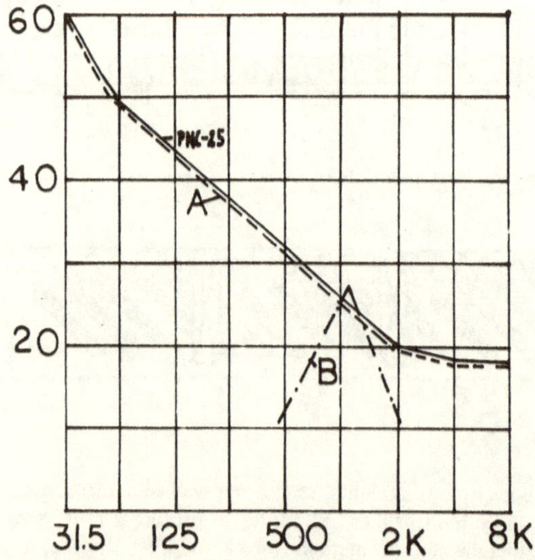

Fig. 15.3 Noise A is rated less than *PNC*-25, has A-weighted sound level of 35 dB-A, and should be considered acceptable; "nonacceptable" noise B exceeds *PNC*-25 but has an A-weighted sound level of only 26 dB-A.

5. *Reverberation time at 500 Hz.* The optimal reverberation period in the mid-frequency of the audio spectrum should be in the order given by the equation (see Fig. 15.4)

$$T = 0.15 \log V - 0.15 \qquad = 0.15 \log V' + 0.07$$

where

V = volume of the control room (ft^3)
V' = volume of the control room (m^3)

Thus, when the enclosure has a volume of 5000 ft^3 (141.5 m^3), the reverberation time should be

$$T = 0.15 \log 5000 - 0.15 \qquad = 0.40 \text{ s}$$

There is a point to be made about reverberation which is not generally

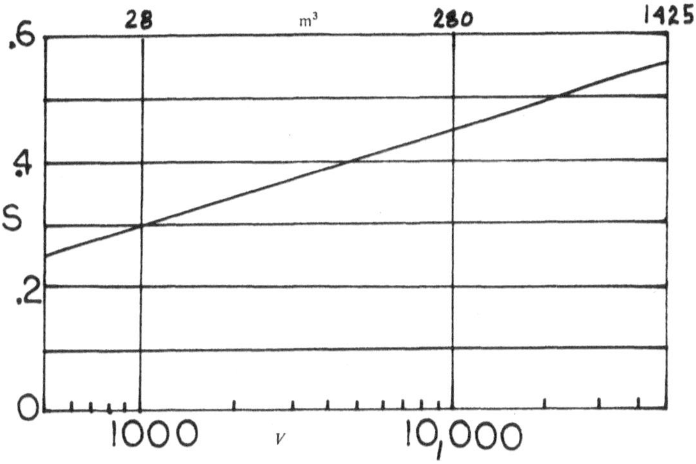

Fig. 15.4 Variation of 500-Hz reverberation time with volume of control room *V* expressed in cubic feet

known. When a listener is close to a source of sound, and the signal is suddenly terminated, the subjective impression of the reverberation is much less than when one is at a greater distance in the room. It has similarly been learned that when one measures the reverberation time in a live room, with the microphone close to the source of sound, there occurs initially in the graphically recorded sound decay a sudden steep drop in the sound level, after which the curve assumes the slope obtained at most other places in the room. The point of the discussion is that when a mixer is too close to the monitoring units, and hears practically no reverberation, he or she may be tempted to introduce a reverberatory note to the recordings which the mixer might not do if he or she sat a little farther from the units.

6. *Loudspeakers.* The purpose of the following is not to recommend certain loudspeakers for the control room, only to indicate that when the frequency response of the recordings is to extend to 30 Hz, few single woofers are capable of providing adequate low-frequency sound pressure levels in it. This may be demonstrated as follows.

Assume that the sound power level requirements of a loudspeaker at 10 ft from the unit is of interest, where the direct sound energy density predominates over the generally reflected one. The pertinent equation for this purpose is

$$
\begin{aligned}
WL &= SPL + 20 \log R - 10 \log Q + 0.7 \\
&= SPL + 20 \log R' - 10 \log Q + 11
\end{aligned}
$$

where

WL = sound power level relative to 10^{-12} W
Q = directivity factor
SPL = sound pressure level relative to 0.0002 μbar
R = distance between loudspeaker and observer (ft)
R' = distance between loudspeaker and observer (m)

When $Q = 2, R = 10$ ft, then

$$
\begin{aligned}
WL &= SPL + 20 \log 10 - 10 \log 2 + 0.3 \\
&= SPL + 17.3
\end{aligned}
$$

In terms of the sound power,

$$W = 10^{(SPL - 102.7)/10}$$

If $SPL = 112.7$ dB at 10 ft the acoustic power at any frequency is 10 W.

The peak displacement of the loudspeaker cone from its mid or rest position, when the unit is mounted in a large baffle and assumed to move as a piston, is given by

$$d = \frac{115{,}200 \quad 10^{(SPL - 102.7)/10} \; {}^{1/2}}{f^2 D^2} \qquad \text{(in.)}$$

$$d' = \frac{1{,}888{,}128 \quad 10^{(SPL - 102.7)/10} \; {}^{1/2}}{f^2 D'^2} \qquad \text{(cm)}$$

where

f = frequency
D = cone diameter (in.)
D' = cone diameter (cm)

If $f = 30$ Hz, $D = 16$ in., $SPL = 112.7$ dB, then

$$d = \frac{115{,}200 \quad 10^{(112.7 - 102.7)/10} \quad ^{1/2}}{30^2\, 16^2}$$

$$= 1.6 \text{ in.}$$

This is far too great an excursion for a single loudspeaker cone. If it is assumed that the maximum travel of a diaphragm is in the order of 0.4 in., four units are required. As one may readily learn by examining the so-called equal loudness contours for the human ear, for a 30-Hz note to sound as loud as a 1000-Hz tone with a sound pressure level of 90 dB, the sound pressure level for the low frequency has to be at least 20 dB higher, so that the figure of 112.7 dB employed above in the calculations is quite realistic.

Note also that it is not the required acoustic power of 10 W which is the critical quantity, since even for a 10 percent efficiency on part of the loudspeaker, the electric power is still only 100 W (not a difficult quantity to achieve in modern power amplifiers), but rather the required cone travel of the woofer.

It is also recommended that the loudspeakers be placed in a large cabinet extending the width of the control room, so as to avoid undesirable diffraction effects produced by the projected panels of the individual loudspeaker housings.

Fig. 15.5 shows the required cone travels for a 16 in. diameter loudspeaker at 30 Hz and at 10 ft for various sound pressure levels.

The cyclic air volume displacement required by a vibrating baffled piston in free air to achieve a given *SPL* at a distance R (ft) or R' (m) at a frequency f comes to

$$V = \frac{0.00004R \times 10^{SPL/20}}{f^2} \text{ ft}^3 \qquad V' = \frac{.0000034R' \times 10^{SPL/20}}{f^2} \text{ m}^3$$

$$= \frac{0.069R \times 10^{SPL/20}}{f^2} \text{ in.}^3 \qquad = \frac{3.4R' \times 10^{SPL/20}}{f^2} \text{ cm}^3$$

Fig. 15.6 shows this relationship graphically for an observer distance of 10 ft. Thus, for an *SPL* of 120 dB at 10 Hz the cyclic volume displacement comes to 4 ft^3 or 0.1 m^3.

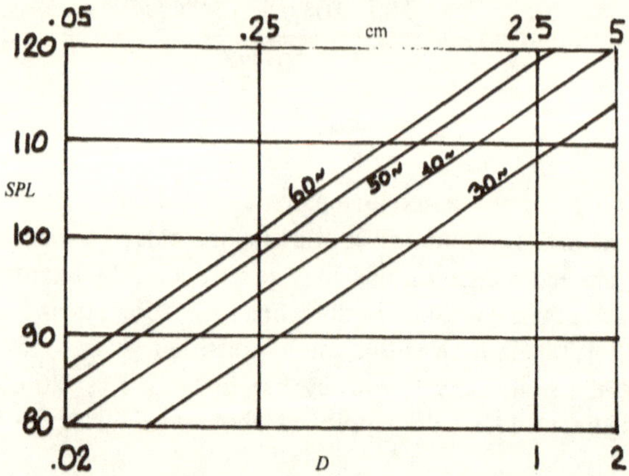

Fig. 15.5 Required travel *D*, in inches, of 16 in. diameter loudspeaker cone from its mid or rest position at different frequencies to generate, at 10 ft (3.05 m), the sound pressure level, *SPL*, shown on the ordinate

For frequencies in the infrasonic range, the required cyclic air volume displacements become exhorbitantly large, and it becomes difficult to construct generators capable of radiating high loudness levels below 15 Hz.

Fig. 15.7 shows the variation of the sound pressure level with the frequency below 100 Hz for various loudness levels in phons. The equation describing the *SPL* as a function of phons and frequency is

$$SPL = 160 - (67.5 - 0.42N) \log f$$

where N is loudness level in phons.

For example, at 10 Hz, the *SPL* corresponding to a loudness level of 120 phons comes to

$$SPL = 160 - (67.5 - 0.42 \times 120) \log 10$$
$$= 142.5 \text{ dB} \qquad \text{(see dot in Figure)}$$

The corresponding cyclic air volume displacement for an *SPL* of 142.5 dB at 10 ft (3 m) is, therefore, 53 ft^3 (1.5 m^3).

A volume displacement of 53 ft^3 is equivalent to one of 1.5×10^6 cm^3. Assume a circular piston of 45 ft or 1382 cm diameter; its area is 1.5×10^6 cm^2. For a volume displacement of 1.5×10^6 cm^3 the travel of such

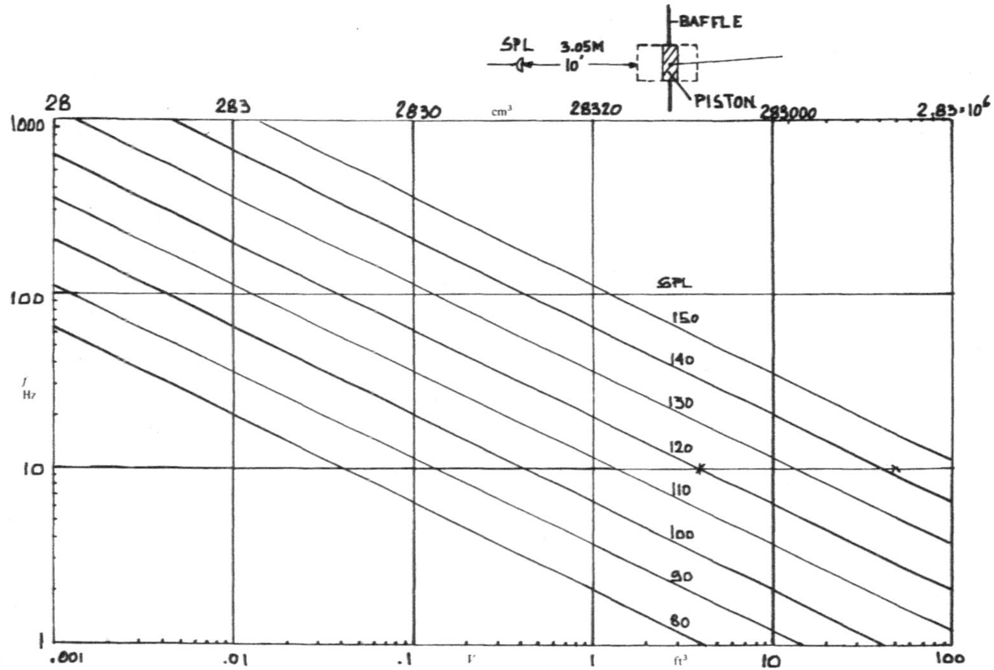

Fig. 15.6 Required cyclic volume displacements for a baffled piston to generate at 10 ft (3.05 m) from an observer the *SPL*'s indicated on the figure. Example: To generate an *SPL* of 120 dB at 10 Hz, and at 10 ft, 4 ft^3 of cyclic volume displacement are necessary.

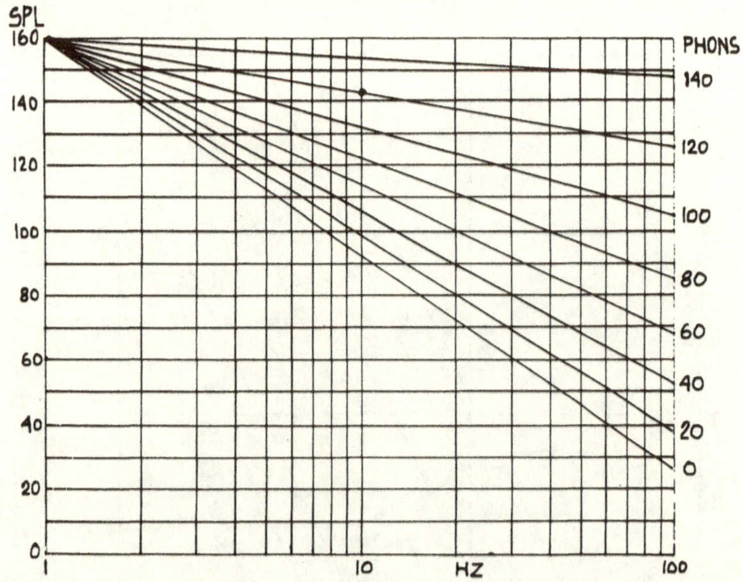

Fig. 15.7 Idealized contours of equal loudness below 100 Hz

a large piston is 1 cm. The air load contributed to the mass of the piston due to the inertia in the air which must be set into motion comes to

$$M = \frac{8R^3\rho}{3} \qquad \rho = \text{density of air} = 0.0012\ \text{g/cm}^3$$

$$= 2.67 \times 0.0012 \times 691^3$$

$$= 10^6\ \text{g}$$

$$\cong 1000\ \text{kg or 1 metric ton!}$$

This is the load when one side of the piston is terminated in an infinite pipe. When both sides of the piston radiate the energy into free space, the load is twice as much, or 2 metric tons! This shows the practical impossibility of reproducing such a subsonic note on an impressive scale in a theater, as in the case of "Sensurround," a low-frequency sound reproduce system employed in theaters for effect. One may obtain a patent for the reproduction of infrasonic notes in show places, so that the tactile or contactual perception of such signals will cause a heightened response to a play or motion picture, but its realization is not possible, economically, in accord with the inventor's idea (see English Patent No. 432,221, granted to George Victor Dowding, July 23, 1935, for the use of infrasonics in theaters).

The reproduction of infrasonic sound would not be fraught with many difficulties if the listener were in a small air-tight enclosure. The "load" on the loudspeaker would be an acoustic capacitance below 10 Hz—the way the air volume between a headphone and the ear comes close to an acoustic capacitance. Thus below 10 Hz only the required volume amplitude is needed and not the radiated power to obtain the required sound pressure level.

7. *Mixing console.* Since this unit is generally placed on a raised platform, a railing or balustrade should be installed around its perimeter so that occasional visitors to this area could not fall off when accidentally stepping sideways.

Also, the metal panels on the sides of the console should be treated on the inside with a viscoelastic damping compound to avoid rattles.

When a parallelism exists between the reflective front wall below the window and any of the panels of the console, sound-absorbent treatment should be applied to one or both of these surfaces to prevent a pronounced resonance condition in front of the unit.

Figs. 15.8 and 15.9 show a recommended plan and elevation of a control room which incorporates the acoustic features discussed above. Note the three-position raised console and behind it the still higher couch for the producer or sponsor. The horizontal symmetry features two doors, one of which may be false, the convex Lucite or Lexan tier drop ceiling splays through spot lights which can be transmitted onto the console, and the anechoic rear wall treatment.

These design details are somewhat contrary to the present fashion of an acoustically large but visually small control room in which a furled curtain is placed about three feet away from the walls. But the present control rooms are already too small, and if they are made still smaller, they will

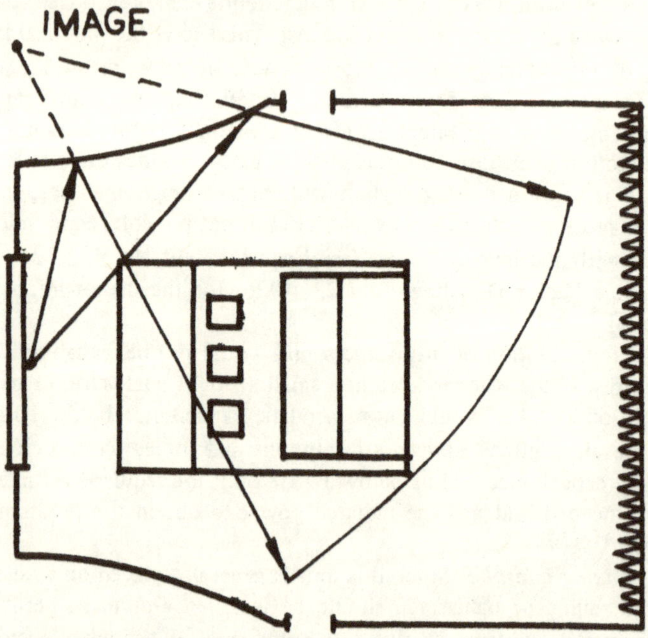

Fig. 15.8 Recommended plan for control room of 5000 ft³

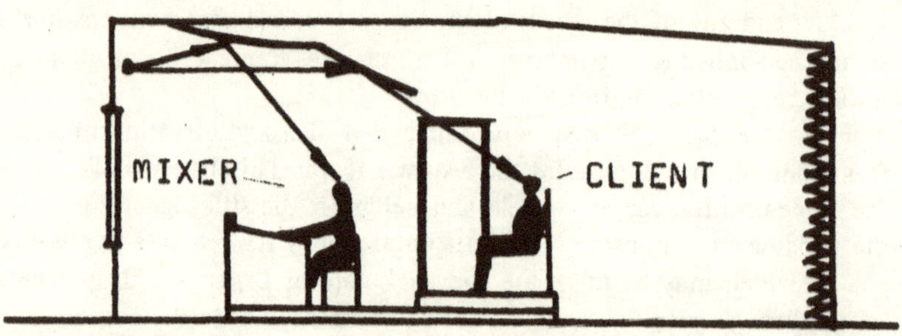

Fig. 15.9 Recommended elevation of control room of 5000 ft³

depart too far from the final reproducing enclosure, the living room or den in the home. Such a little enclosure provides neither aural nor psychological correlation between mix and ultimate reproduction.

Sound-retardant control room windows larger than 1 m × 3 m (3.28 ft × 9.85 ft) may produce undesirable sound reflections in the recording studio in which the program is monitored in the control room. The effect is most pronounced for wavelengths the equal to or smaller than the size of the window, and therefore generally concerns frequencies which are above 500 Hz. The same situation prevails in re-recording studios where the mixing console may be a short distance from the rear wall which may contain large windows for either a client's room, a voice booth (where an actor may watch the action on the screen for which he is to add a voice track in the case when the "looping" in another enclosure may not be practical), or where an acutal control room is located because the dubbing stage is a multiple-purpose facility.

To employ ordinary glass for a convex window pane is not only undesirable from a cost point of view in that it would have to be specially formed in the factory, but is undesirable also from a breakage point of view if the pane is thick. For this reason curved acrylic sheets should be installed in the window opening, because they are nearly shatterproof. Plexiglas, Lucite, and other such transparent sheets may be used if they are 6 mm (¼ in.) thick so that they may be "cold-bent." Polycarbonate material weighs 7.1 kg/m^2 when 6 mm thick (1.42 lb/ft^2 when ¼ in.-thick), or about half as much as glass, and has a tensile strength of 6.3×10^6 kg/m^2 (9000 $lb/in.^2$), and when cold-bent introduces no visible distortion in the sheet.

Acrylic resins are thermoplastic materials of high light-transparency which are frequently sold under such tradenames as Plexiglas and Tuffak (Rohm and Haas Company) and Lucite and Lexon (E. I. Du Pont de Nemours & Co.). Plexiglas and Lucite have a light transmittance of 0.92, compared to 0.89 for ½ in. plate glass, while Tuffak and Lexon have a light transmittance of only 0.82 to 0.89. The latter two plastic materials are not quite as good optically as the former, because they are extruded while Plexiglas G and Lucite AR are cell-cast. Although Tuffak and Lexon exhibit greater impact strength than Plexiglas and Lucite, nevertheless the latter two have from 6 to 17 times more impact resistance than double-strength glass.

The reason why the reflections from a convex window on the studio side are generally innocuous compared to those from a plane window is that the wavefronts of the convex window reflections are greater than those from a flat hard plane when the distances from the source to the window and from the window to the observation point in the studio are the same in these two cases.

In the following the graphic construction of the reflected wavefront is based on the analogous optical case, because it lends itself to relatively simple geometric constructions. However, as shown in Fig. 15.10, when the reflected wavefront is constructed by making the angle of reflection, *r*, equal to the angle of incidence, the length of the wavefront *m'* is actually longer than the length of the reflected wavefront constructed by employing *V* distance for the center of its circle.

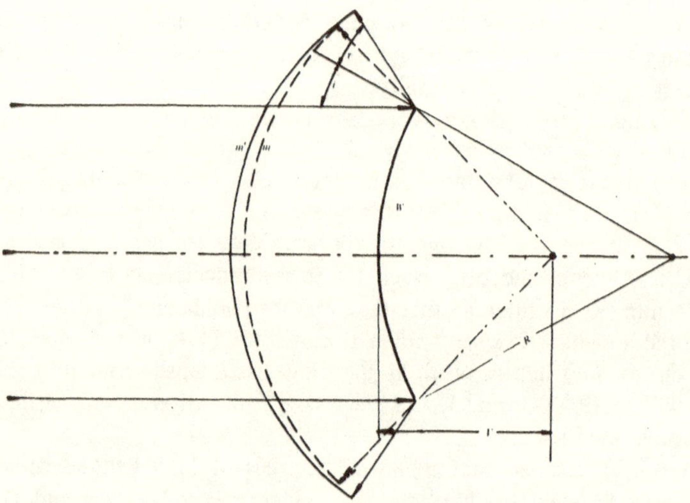

Fig. 15.10 The solid line *m'* represents the geometrically constructed wavefront produced by the circularly convex window *W*, whereby the angle of reflection *r* is equal to the angle of incidence *i*. The dashed line *m* represents the wavefront constructed by using the distance *V* for the center of the wavefront circle.

Fig. 15.11 shows a graphical representation of the reflection of sound from a circularly convex window, taken from the analogous optical case, which is valid for small central reflecting zones and surfaces that are large compared to the incident wavelength. It also illustrates the magnification, m/l, of the wavefront from a convex window over that from a flat window of equal width *W*. In the following example it is assumed that the curved length of the window is equal to radius *R*, so that the included angle equals 1 radian or 57.3 degrees.

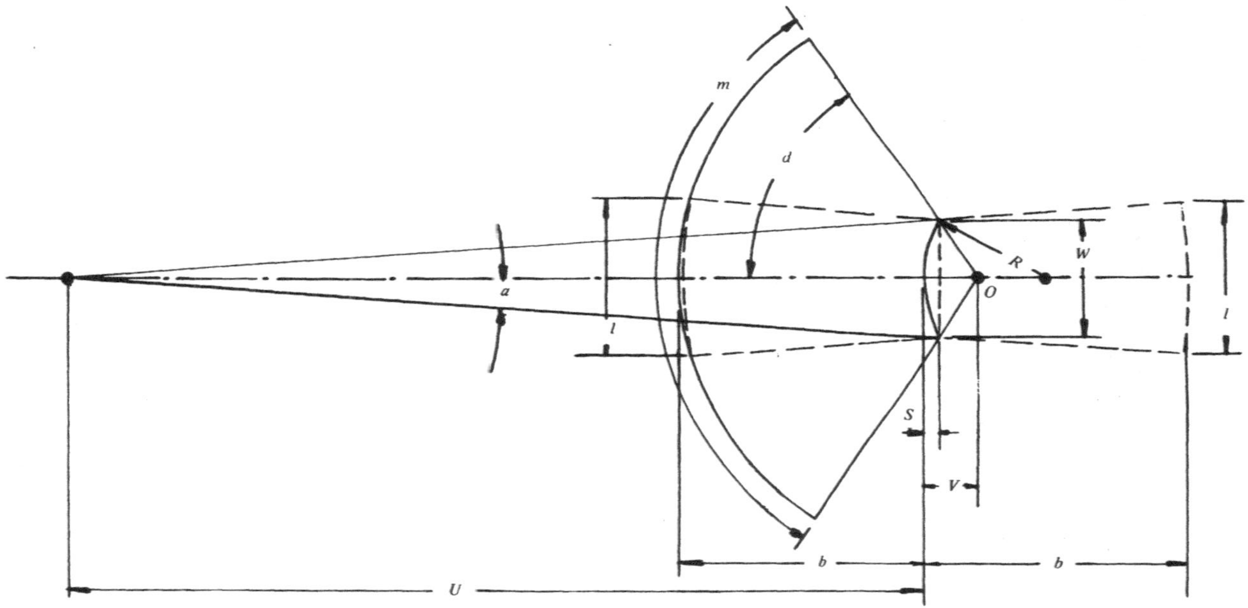

Fig. 15.11 Construction of reflected wavefront produced by a circularly convex window. W: width of convex or flat window; b: observer distance; R: radius of convex window; U: distance from source to window; V: distance between window and point O, which is the center of the circular wavefront reflected from the convex window; l: length of wavefront reflected from flat window at distance b; m: length of wavefront from convex window; S: sagitta.

Let

U = 70 ft (source distance, between source and convex window)

R = 10 ft (radius of convex window)

W = $2R \sin \times (57.3/2)$ or 9.588 ft (horizontal width of window)

b = 20 ft (observer distance, between observer and convex window

V = image distance, that is, the distance from the convex window, measured along the longitudinal axis of the system, to the center of the circle describing the circular wavefront

$$= \frac{UR}{2U+R} = \frac{70 \times 10}{140+10} = 4.67 \text{ ft}$$

$$= \frac{R}{2} \quad \text{if } U \text{ is infinity}$$

S = sagitta, the distance between midpoint of arc to midpoint of chord

$$= R[1 - \cos(57.3/2)]$$

$$= 10(1 - 0.877) = 1.225 \text{ ft} = 14.7 \text{ in.}$$

$$d = \tan^{-1} \frac{0.5W}{V-S} = \tan^{-1} \frac{0.5 \times 9.588}{4.65 - 1.225} = 0.9477 \text{ radians}$$

$$= 54.3 \text{ degrees}$$

$$a = \tan^{-1} \frac{0.5W}{U+S} = \tan^{-1} \frac{0.5 \times 9.588}{70 + 1.225} = 0.0673 \text{ radians}$$

$$= 3.85 \text{ degrees}$$

m = length of wavefront reflected from convex window

$$= 2(V+b)d$$

$$= 2(4.67 + 20) \times 0.9477 = 46.76 \text{ ft}$$

$$
\begin{aligned}
l &= \text{length of wavefront reflected from flat window} \\
&= 2(U + b)a \\
&= 2(70 + 20) \times 0.0673 = 12.11 \text{ ft}
\end{aligned}
$$

$$
\begin{aligned}
\frac{m}{l} &= \text{magnification of wavefront reflected from convex window} \\
&= \frac{47.76}{12.11} = 3.86
\end{aligned}
$$

This means that the energy density of the reflected wavefront from the convex window is almost one-fourth that of a flat window of the same size, or 10 log 3.86 ≅ 6 dB less. The corresponding sound pressure level reduction is 12 dB less, because the energy density is proportional to the square of the sound pressure. Hence the effect of the sound reflection from the circularly convex window is far less objectionable 20 ft from the window (the position of the mixing console in the re-recording studio) than the sound reflection from a flat window of the same size, with the source distance and observer distance remaining equal.

For a spherically convex window the wavefront magnification would have been equal to $(m/l)^2$ when the window was square.

Because plastic sheets come in standard sizes, like 2 ft × 4 ft, 4 ft × 6 ft, 6 ft × 8 ft, 8 ft × 10 ft, 10 ft × 12 ft, (¼ in. thick) it is advisable to choose the window area in accordance with the available panel dimensions. A convenient design for such a circularly convex window is to make the curved portion of the window equal to 4 ft, 6 ft, 8 ft, 10 ft, etc., and also to make the radius of curvature R equal to the curved panel length. This makes the straight or horizontal width of the window W equal to $0.96R$, and the included angle equal to 57.3 degrees or 1 radian. The depth of the curved portion, the sagitta S, which may be defined as the distance from the midpoint of the chord to the midpoint of the arc, is given as

$$
\begin{aligned}
S &= R(1 - \cos 28.65°) \\
&= 0.122R \text{ ft} \\
&= 1.47R \text{ in.}
\end{aligned}
$$

By this method of design, the following window dimensions result:

R (ft)	*S* (in.)	*W* (ft)
2	2.94	1.92
4	5.88	3.84
6	8.82	5.76
8	11.76	7.68
10	14.70	9.60
12	17.64	11.52
14	20.58	13.44
16	23.62	15.36
18	26.46	17.28
20	29.40	19.20

It is also possible to employ a convex window inside a control room, although generally it is only reflected sound that can strike the window, because the loudspeakers are above it, as shown in Fig. 15.12.

The reflected wavefront from a convex window can be made still greater when the window is made spherically convex, although this may result in considerable constructional difficulties.

Sound-retardant windows with one or two external convex panes of ¼ in.-thick acrylic material generally require one or two internal glass panes when the window is to exhibit a sound transmission loss of 50 dB at 500 Hz. Thus, with one ¼ in.-thick Plexiglas sheet with a curved length of 10 ft, one 3/8 in.-thick glass pane, and another, 1/2 in.-thick glass pane separated from the first by an average of 8 in., is necessary to achieve the desired sound attenuation for the window at 500 Hz, as shown in Fig. 15.13.

There is no adequate substitute for a curvilinearly convex window to provide a uniform distribution of the reflected sound field. This is illustrated in Fig. 15.14, where flat angled windows are used instead of a circularly convex one. There are "acoustic shadows" in the reflected sound field, which may be more objectionable than the reflected sound field created by a single flat window.

It is important to understand the advantages of a reflective environment around the loudspeakers in a control room. Attempts have been made to introduce sound absorption around such radiators to minimize acoustic comb filter effects when first-order reflections about the radiators meet with the direct sound wave at the mixing console.

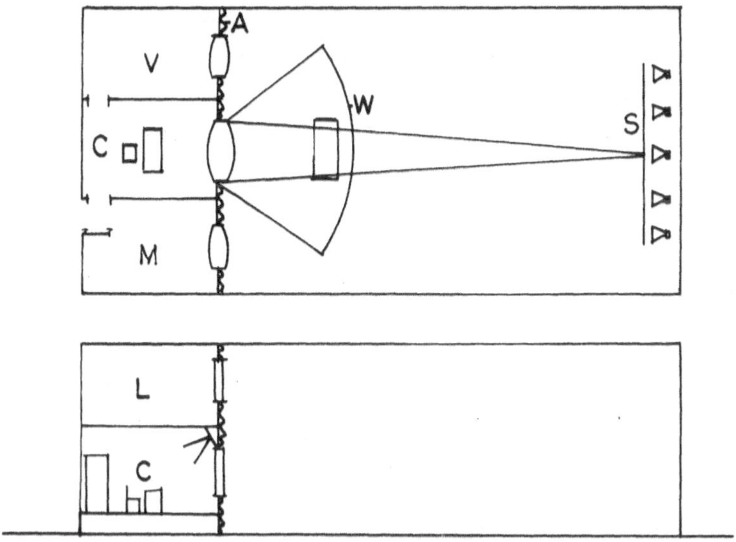

Fig. 15.12 Plan and elevation drawing of a rerecording studio with convex windows. V: vocal booth; C: control room; M: machine room; S: screen; A: absorbent material; L: lounge; W: wavefront reflected from convex window.

Interference produced by a single reflection and a direct sound ray can readily be depicted graphically, and a real-time analyzer can also demonstrate this phenomenon in the open when a single microphone is employed. It is more difficult to calculate the quantitative effect and to evaluate the qualitative effect of more than one first-order reflection for a binaural receiver. A dummy head with a high-quality condenser microphone in each ear, a phase-correcting network for the transducers, rectification of each microphone before mixing for an average effect, and adequate filtering result in a loudspeaker response curve somewhat different from that obtained for the single receiver in the open with only one microphone. The response irregularities of the loudspeaker tested with several like signals depend on the distance between the transducers, the time intervals between them and their phase relationships. The difficult part of the experiment, however, is to evaluate the qualitative results on music with its highly transitory signals when such first-order reflection interference takes place in an enclosure.

The shorter the time intervals between the individual reflections and the direct sound the less is the ear able to detect acoustic comb filter effects. Also, such reinforced music is often evaluated as more pleasing specifically for its variety of phase effects. By the same token, the music of many violins in a symphony orchestra is preferred over that of a few such string instruments, even though there exists a wide variety of phase relationships between the many violins. And what mixer has not noticed the unpleasant reverberatoriness resulting from a "one-dimensional" reverberator like a spring compared to the pleasing impression achieved with a "three-dimensional" unit like a chamber when "reverb" is added to a program?

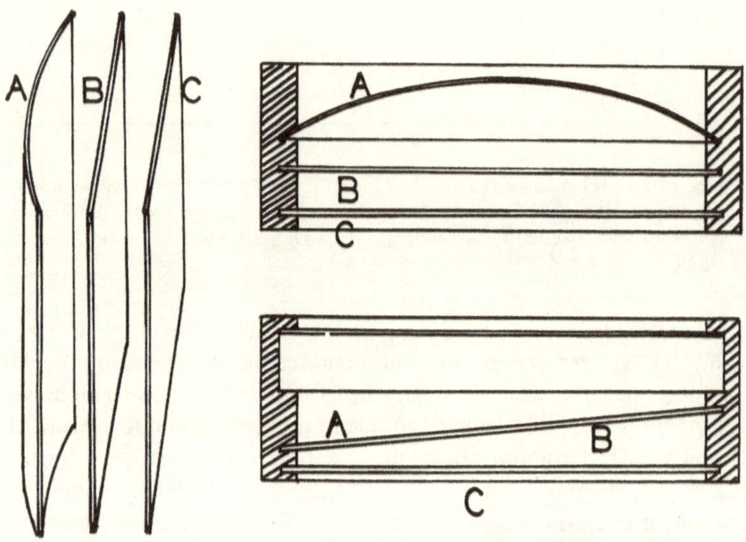

Fig. 15.13 Plan, elevation, and perspective view of a triple-pane, sound-retardant convex window. A: ¼ in. Plexiglass; B: 3/8 in. glass; C: ½ in. glass

Fig. 15.15 shows the directional characteristics of a vibrating piston in an infinite baffle for various ratios of D/λ, the ratio of piston diameter to the wavelength of sound. For such a device

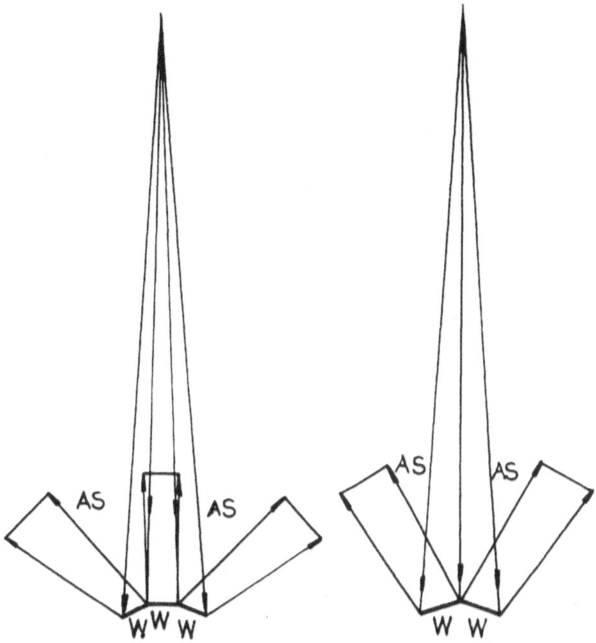

Fig. 15.14 Diagrams show that when 2-segmental or 3-segmental windows, W, with straight panes are used instead of curvilinearly convex ones, acoustic shadows, AS, are created in the reflected sound field

$$R = \text{Response} = \frac{J_1 \dfrac{kD}{2} \sin a}{\dfrac{kD}{4} \sin a}$$

where

J_1 = Bessel function of the first order
D = diameter of piston
k = $6.28/\lambda$
a = angle of radiation
λ = wavelength of sound

When the parameter D/λ is multiplied by the speed of sound divided by D the frequency of the signal radiated by the piston is obtained.

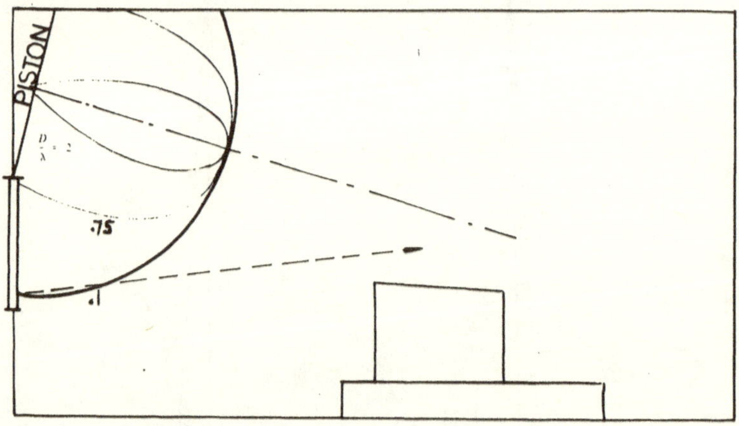

Fig. 15.15 Directional radiation characteristics of a circular piston in an infinite baffle for the various ratios of the piston diameter to the wavelength of the generated sound.

The previous equation for the directional characteristics of a vibrating circular piston set in an infinite baffle may be approximated by the formula

$$R' = \frac{X^2 - 16}{X^2 + 16}$$

where $X = 3.14D \sin a/\lambda$.

Fig. 15.16 shows the relative responses by both equations, and it is seen that the deviation of R' is not large compared to R.

To illustrate the undesirable relationship of placing a loudspeaker in an absorbent front wall of a control room where there is a large window below the radiator, consider a woofer with a 1.5 foot diameter diaphragm emitting a 100-Hz tone, so that D/λ comes to 0.13. For such a unit the directional output is practically a hemisphere. At 90 degrees, the response is 0.96, or 96 percent of the output along the normal or longitudinal direction. This means that almost as much sound is directed at the control room window as is directed to the mixer. As shown by the dashed line in Fig. 15.17 a strong first-order 100-Hz reflection is received by the mixer—without any first-order mid-range or treble reflections when the environment about the loudspeaker is rendered highly sound-absorbent. This

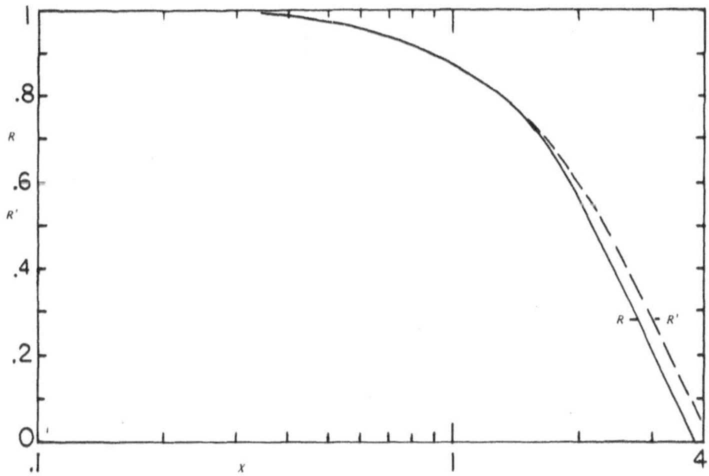

Fig. 15.16 Percent sound pressure response of a vibrating circular piston set in an infinite baffle. R is the response determined by the equation $J_1(X)/0.5X$; R' by the equation $(X^2 - 16)/(X^2 + 16)$. $X = 3.14D \sin a/\lambda$, where D is piston diameter; a is the radiation angle, λ is the wavelength, and J_1 is the Bessel function of the first kind.

results in an amplitude distortion of the signal not unlike that produced by raising the low-frequency response of the reproduce amplifier. Very likely, the mixer in such a situation is tempted to attenuate the bass notes of his program, to the disadvantage of the final listener in his living room or den.

To minimize excessive time intervals between first-order reflections and the direct sound at the mixing console, reflective panels should be placed around the loudspeakers in the control room, as illustrated in Fig. 15.17, chiefly for the high notes. There is presently no way to make a window sound-absorptive to reduce the time lag of first-order bass reflections.

Besides the qualitative disadvantages connected with an absorbent environment about the control room loudspeakers, there are also other objections to such a configuration. Since it would obviously not be desirable to have a dead front wall both in front of and behind the mixer, the new concept requires a live surface behind the console. This leads to other problems, particularly in respect to the shape of this reflective rear surface. To prevent echoes, it has to be deeply corrugated so that the low-frequencies are adequately dispersed so that a standing wave phenomenon does

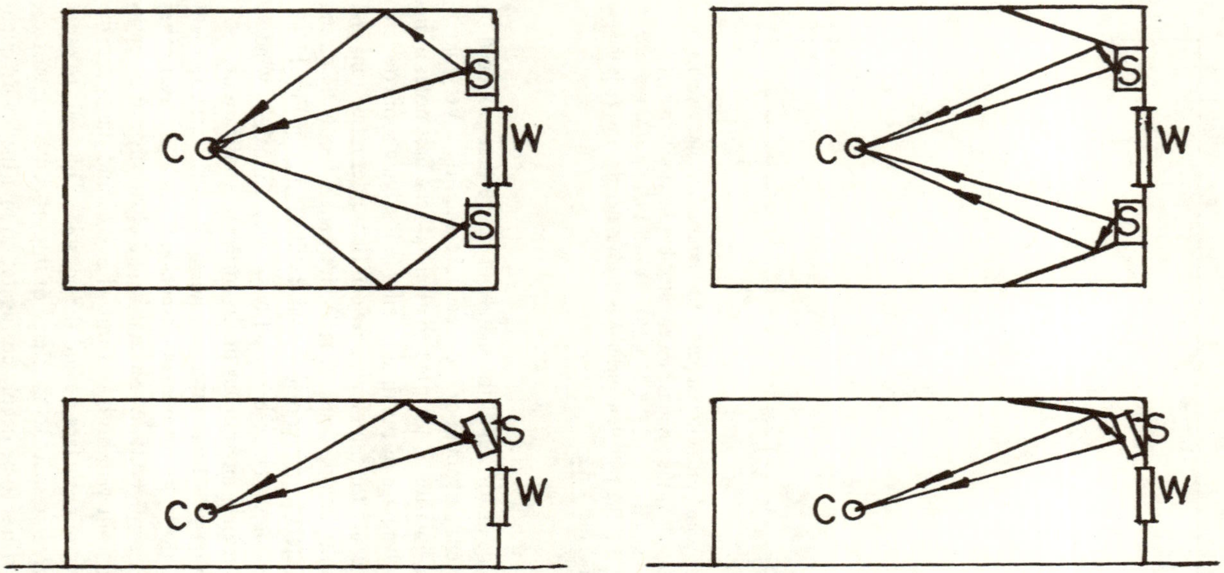

Fig. 15.17 The plan and elevation of the control room shown on the left side of the figure illustrate the relatively long intervals of time between first-order reflections when the walls and ceiling of the room are utilized as reflectors; those on the right show the resulting short intervals of time when reflectors are placed about the loudspeakers. C: console; S: speaker.

not result. It may lead to longer time intervals between the first reflection from that wall and the direct sound, because of the distances involved between console and wall.

When a loudspeaker operates in the open or in an absorptive environment about itself, an acoustic impedance mismatch results between the unit and the air. When reflective surfaces are near the radiator (particularly when the unit is placed at the throat of a horn), this hard environment acts as an acoustic transformer to achieve a higher power conversion efficiency for the system.

Some people may like an acoustically inverted control room just as a difference in preference exists between rock-and-roll and classical music.

When this writer visited a LEDE control room, the first thing I became conscious of, even before music reproduction started in the room, was the reflected tape machine noise from the hard rear wall, near which so much recording equipment is presently located. By simple analysis, whatever noise exists at the console when the rear wall is absorbent, will be 3 dB higher when the wall is made reflective. In the particular control room, with a rear wall corrugated to a depth of only 3 in., the noise level at the console was objectionally high, and efforts were made to hang a curtain over the machines during a "take"–obviously a Rube Goldberg solution to the problem. An uneven reproduction level was also apparent behind the console as one moved laterally across the enclosure. On the whole, the quality of the reproduced music did not seem to be superior in this type of monitoring enclosure.

It should be noted that with control rooms being as dead as they are, it is practically useless to speak about a reverberant field about the console. Since the significant sound decay occurs only during the first 10 dB, this is equivalent to only 0.050 s in an enclosure with a reverberation time of 0.3 s. Hence, the reverberation period in a control room is generally not significant because of its brevity. It is, therefore, difficult to form a judgment about the music quality in such an enclosure, whether or not the front wall or the rear wall is reflective.

Chapter 16

ADR Studios

Looping or ADR (Automatic Dialogue Replacement) rooms have also been called post-synchronizing studios because they provide acoustic signals after photography. The term looping derives from the fact that film loops are used in the projection of scenes for which speech has to be recorded and, for one reason or another, could not be recorded satisfactorily at the time the scene was photographed. One such reason might have been the presence of unavoidable noise, like street traffic, seaside surf, etc., or because the actor or actress had laryngitis or another ailment which precluded the formation of intelligible speech. The term automatic dialogue replacement springs from the fact that a number of operations (reproduce, record, erase, projection, etc.) are controlled by automatic equipment, so that a scene may be reset to any position, as indicated for instance by a film footage counter.

In practice, a scene is projected on a motion picture screen and the actor watches the scene while listening to his cue (the original unsatisfactory sound track) over a pair of earphones for pacing. He tries to repeat his lines in front of a microphone in synchronism with his lip movements on the screen. When a satisfactory recording has been made, the film in the projector is advanced to another scene, for a similar operation.

"In the old days" a sound editor had to prepare film and dialogue loops by cutting the pertinent scenes from the roll of picture film and the pertinent cue dialogue from its sound reel, and forming loops of film which could be reproduced repeatedly.

A sound mixer is located in a separate room and watches both the projected scene and the actor through a sound-retardant window. He monitors the rehearsals and operates the equipment which records the rehearsal dialogue as well as the final takes. He is also in control of the auto-looping system, that is, the bidirectional interlock film transport and the "Start," "Stop," "Look," "Ready," etc., operations. The sound is generally recorded on a three-track 35 mm magnetic film operated by a remotely controlled film transport unit, although magnetic tape may also be used.

Because the locale of the scene may vary from outdoors to reverberant churches, such a studio should exhibit varying reverberatoriness in order to make possible the correct acoustic simulation of the projected locale. This is accomplished by mounting on the walls near the screen vertically hinged panels which are highly reflective on one side and highly absorbent on the other, so that by turning them the environment of the microphone can be made either "live" or "dead."

Such rooms permit also the recording of sound effects, like tap dancing, scuffles, splashings, etc., wherefore they often contain pits filled with sand, gravel, water, etc., called Foley pits, after an MGM engineer of the early 1930's who first used such containers for the production of sound effects.

This studio is highly quiet-demanding, and the air velocity of the air-conditioning system should not be more than 500 ft/min.

Fig. 16.1 shows a plan of a recently constructed post-synchronizing studio. Note that both video tape and film can be reproduced in the room on the large screen. The vocal booth adjoining the mixing room is for the introduction of nonsynchronous, off-stage dialogue by an actor or actress appearing in the feature.

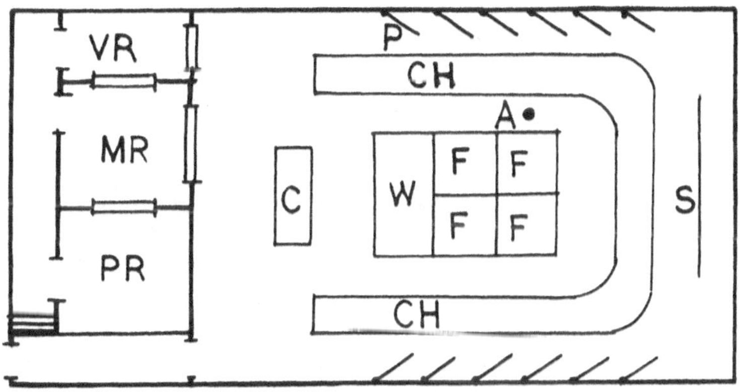

Fig. 16.1 Post-synchronizing or ADR (Automatic Dialogue Replacement) studio. VR: vocal room; MR: mixing room; P: hinged sidewall panels which are reflective on one side and absorbent on the other; W: wood floor for the production of the clicks in tap dancing; CH: 10 in. (25 cm) deep channel which can be filled with water or used as a walkway; F: Foley pits for the generation of sound effects; S: screen; PR: projection room for film or tape; C: console; A: actor.

Chapter 17

Re-Recording Studios

A re-recording or dubbing studio is a room for combining sound tracks with picture film. The sound tracks may be effects tracks, like the slamming of a door, the squealing of automobile tires, winds, thunder, etc., or they may consist of musical passages, vocal or instrumental, which have to be added to the picture for realism or "mood." Sometimes the voice tracks of an actor or actress have to be "equalized" for better intelligibility, or an entire new voice track has to be added for any number of reasons.

Frequently the room is also employed as a review room, where the performers and executives may view a picture.

Somewhere in the rear part of the studio the rerecording console is located which is required to mix the many voice, effects, and music tracks to best advantage for the feature.

In the case of multi-sound tracks, like Cinemascope, Todd-AO, and others, the reproduction of the signals requires several loudspeaker systems behind the screen, and a number of surrounding loudspeakers on the sidewalls or the ceiling. For the reproduction of "Sensurround," at least four very large loudspeaker enclosures have to be installed in the boundaries of the studio to allow the reproduction of high-amplitude low-frequency signals.

Such a critical room, obviously, should not exhibit acoustic defects like echoes, sound concentrations, delayed reflections, or a prolonged reverberation.

For a highly attractive interior decor, carpet should be installed on the floor, the lighting should be recessed, and the velocity of the air of the air-conditioning system should nto be higher than 500 ft/min to reduce turbulency effects at the grilles.

For the best acoustic effect, the sidewalls of the re-recording studio should be splayed for the first three-fifths of the length of the room, with the treatment starting at the screen. The splays should not be carried behind the screen, where the acoustic treatment should consist of 1 in. thick Owens-Corning black duct liner (black for maximum picture contrast on the screen).

The function of the splays is twofold: to avoid a parallelism (which could give rise to echoes) between the sidewalls near the screen, where the surfaces should be reflective, and to direct a sufficient number of first-order sound rays to the mixing console at the rear of the enclosure to assure adequate sound reinforcement for presence there. This is accomplished by making the triangular corrugations reflective, and slanting them 1 ft in 8 ft, which comprises an angle of 7 degrees; and by making the reverse parts of the splays sound-absorbent so as not to reflect sound back to the screen.

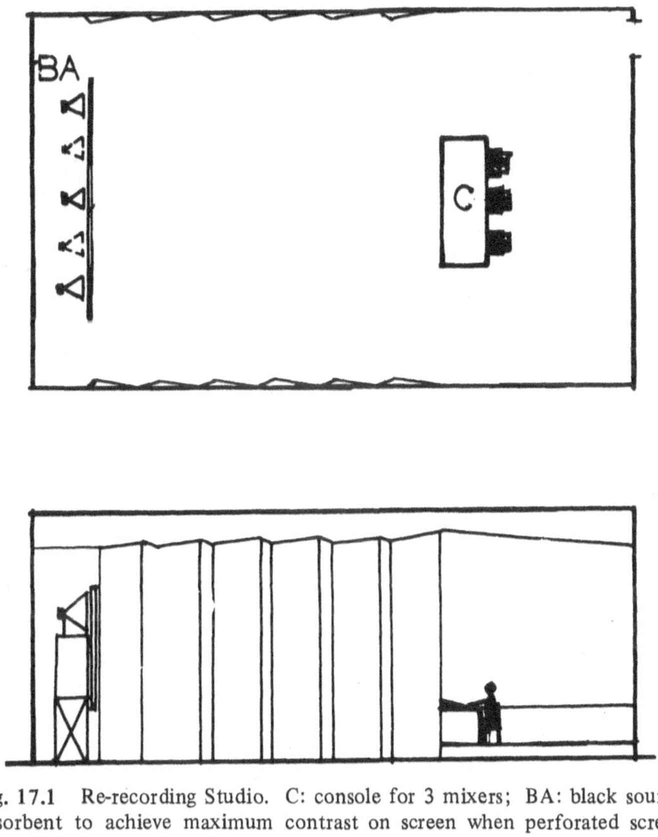

Fig. 17.1 Re-recording Studio. C: console for 3 mixers; BA: black sound-absorbent to achieve maximum contrast on screen when perforated screen allows light to pass through it.

The rear portions of the sidewalls should carry tan, buff, gray, etc., sound-absorbent treatment (white would diminish the contrast on the screen too much), installed from the floor to the ceiling, with the lower four feet made of hard and durable Tectum, a National Gypsum Co. product, which can be painted repeatedly without losing its absorptivity.

The rear wall of the studio, behind which the projection booth is located, should receive the same acoustic treatment as the rear part of the sidewalls, as shown in Fig. 17.1.

The ceiling of the studio should be of the suspended type, so that lighting fixtures, air-conditioning system ducts, conduits, etc., can be hidden. The splayed sidewall treatment should be carried across the ceiling in a compatible and congruous manner.

The color of the walls and ceiling should also not be white, but tan or gray, so as not to reduce the picture contrast on the screen. For this reason acoustic material should be purchased prepainted, so that subsequent painting may not reduce its absorptivity. When the product is colored in the factory, it generally is dyed a certain color, or the fibers of glass or wool are colored before they are pressed into a panel, board, or tile. When the acoustic material is soft, a coat of water-color latex paint has often little effect on the absorptivity of the material, as long as the product's "give" or resiliency is not impaired by the paint coat. A thick coat of lacquer or oil paint usually reduces the material to a hard drum skin, without much absorptivity over the entire audio frequency range. Here as elsewhere in acoustic work, an ounce of prevention is worth a pound of cure; an even better saying is "it is much cheaper to do the job right in the first place."

Chapter 18

Reverberation Chambers

18.1 LARGE CHAMBERS

Very live rooms used in sound recording to introduce a reverberatory note into a program material (so-called reverberation chambers) are best built of concrete. This may be poured concrete, concrete tilt-up panels, or 8 or 12 in. concrete blocks. The latter material, by itself, does not exhibit the high sound reflectivity that is necessary for the purpose. This is due to a certain amount of porosity which the blocks have, and also because the mortar joints are rarely perfect, sometimes showing cracks through which cigarette smoke can be blown. Also, many such blocks are insufficiently cured and crack after construction. It is, therefore, necessary that the inside of a reverberation chamber be plastered with concrete plaster and well trowelled to a fine finish. After the material has dried, a coat of water-base latex or polyurethane should be applied to the interior surfaces to seal all pores.

It is possible to construct a reverberation chamber with stud walls and apply a 1 in. (2.5 cm) thick coat of concrete plaster to the interior surfaces of the room. Because of the relatively small surface density of such a wall, that is, its mass per square foot, the boundary can not exhibit the high sound transmission loss generally required for the walls and ceiling, unless a double-stud construction is adopted which employs another 1 in. (2.5 cm) thick concrete plaster on the outside of the room, with a generous amount of sound-absorbent material installed in the interspace between the studs.

Fig. 18.1 shows the cross section of the preferred construction of such a chamber, which should not be called an echo chamber, because echoes are the last things such an enclosure should exhibit.

Note that the concrete floor of the room is not tied into the walls or the foundation, but floats on a cushion of sand. Around its perimeter the

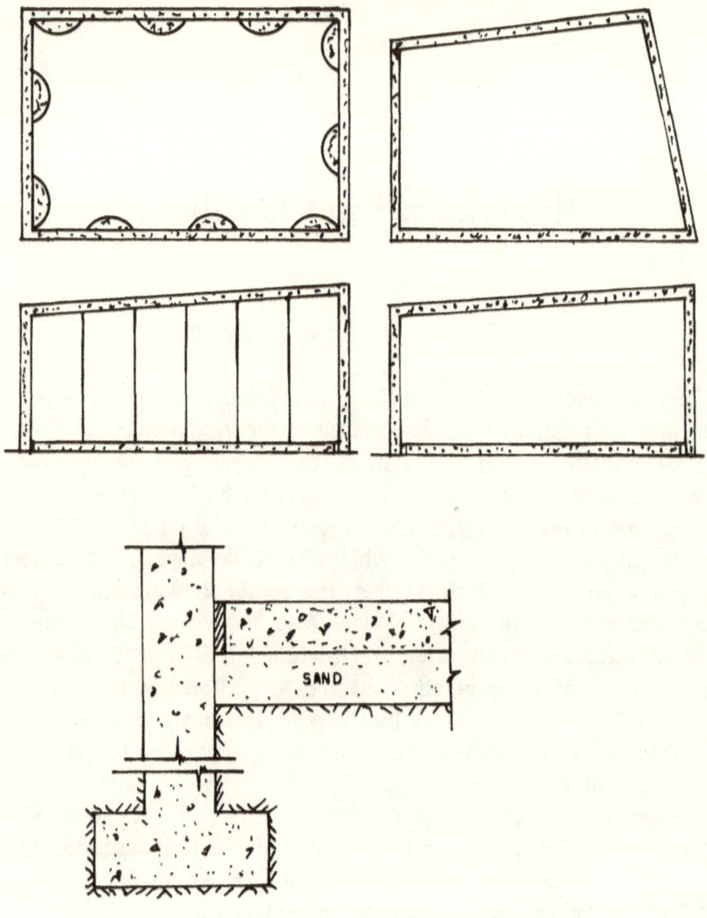

Fig. 18.1 Details of construction, plans and elevations for a reverberation chamber with concrete floor resting on sand.

concrete floor is in contact with a 1 in. (2.5 cm) thick elastic layer of adequate compressive strength to assure structural isolation in case of an earthquake. The latter quantity, depending on the floor area, may be as high as 100 lb/in.^{2} (7000 kg/m^2).

The roof of such a chamber may be built of Spancrete or other such preformed concrete panels.

Fig. 18.1 shows two plans on top of the figure—a symmetrical one to the left and a nonsymmetrical one to the right. Neither has parallel walls and a parallel floor and ceiling. The nonsymmetrical one is probably a little cheaper to construct, although much depends on the volume of the room, its location, and the manner in which it is intended to be constructed.

When the structural requirements of the location are such that the foundations below the floating concrete floor on the grade must be tied together, a "bond beam" is necessary. It consists of bundles of steel bars in a concrete beam which hold the foundations together in case of an earthquake, ground softening, or other hazardous event.

Regarding the sound insulation of a reverberation chamber, NR is defined as the noise reduction of the room, that is, the difference between the outside and the inside sound pressure level at a given frequency. When all the boundaries of a room have the same sound transmission loss TL at a given frequency, and the average absorptivity of the interior surfaces at the same frequency is a, then

$$NR = TL + 10 \log a$$

Thus, for a reverberation chamber with 6 in. (15 cm) thick concrete walls with a TL of 50 dB at 500 Hz and an average absorptivity of 0.01 at 500 Hz

$$\begin{aligned} NR &= 50 + 10 \log 0.01 \\ &= 30 \text{ dB} \end{aligned}$$

But the same value of 30 dB results when $TL = 33$ dB and $a = 0.5$, which may be a room built of 1 in. (2.5 cm) thick fiberboard, according to the equation

$$NR = 33 + 10 \log 0.5$$

The point is that when the inside absorptivity is only 0.01, an outside signal level of 80 dB at 500 Hz appears on the inside of the reverberation chamber as $80 - 30 = 50$ dB which is a high noise level. Even if 12 in. (30 cm) thick concrete were used for the boundaries, which has a TL of 56 dB at 500 Hz, the inside noise level would only be 6 dB less, or 44 dB. Hence it is advisable to locate a reverberation chamber in as quiet a place as possible.

18.2 SMALL AND MINIATURE REVERBERATION CHAMBERS

Because of the high cost of building large reverberation chambers in urban areas where space is at a premium, many devices have been introduced in the recording industry to simulate the reverberatory process in recording. Probably the cheapest is the "one-dimensional" reverberator known as a spring reverberator. It is essentially a long steel spring which is actuated at one end by a vibrator (like a loudspeaker coil) along which the audio signal travels and is picked up at the far end by a proper transducer. There are also "two-dimensional" reverberators, like the Swiss EMT units, which employ a steel plate instead of a spring.

Lately, multi-tapped delay lines of the analog or digital type have been used for a similar purpose. There are also tape reproducers with closely spaced reproduce heads which pick up the recorded signal at discrete intervals, like small echoes. This type of reverberator can, of course, only simulate a decay process but not one of sound growth. As is well known, sound in a reverberant room does not rise immediately to its maximum sound power level, but must increase or "grow" to that magnitude. The situation is not unlike the growth of a direct current in an electrical circuit containing an inductance and a resistance. The "acoustical time constant" comes to $4V/Ac$, where V is the volume of the room, A is the total absorption in the room, and c is the velocity of sound in the air. The equivalent electrical time constant is L/R, where L is the inductance in the circuit and R is the resistance.

18.3 MINIATURE CHAMBER

Regularly over the past 17 years this writer has been contacted several times every year by persons interested in his patent of a miniature reverberation chamber. The patent, which expired on May 30, 1978, is numbered 2,986,228, and is assigned to the Radio Corporation of America, now RCA Corporation.

The device is based on the idea that an enclosure filled with a heavy gas, in which the sound velocity is slower than in air, has a longer reverberation time than an equally large room filled with air, the time being inversely proportional to the speed of sound in the gaseous medium with which the room is filled.

The pertinent equation describing this fact is the Sabine formula

$$T = \frac{13.8 \times 4V}{cSa} = \frac{13.8 \times 4V'}{c'S'a}$$

where

T = reverberation time (s)
c = velocity of sound in the gas (ft)
= 1130 ft/s at normal temperature
c' = velocity of sound in the gas (m)
= 344 m/s
S = total interior surface of room (ft^2)
S' = total interior surface of room (m^2)
a = mean absorptivity of the boundary material
V = volume of enclosure (ft^3)
V' = volume of enclosure (m^3)

The reverberation time of a room is defined as the time that is required for the mean squared sound pressure therein, originally in a steady state, to decrease 60 dB after the source is stopped.

The velocity of sound in meters per second is given by

$$c = \left(\frac{gP_o}{d}\right)^{1/2}$$

where

g = ratio of specific heat of gas at constant pressure to that at constant volume. For practically all gases at normal temperatures this ratio varies little, being in the range of 1.2 to 1.5.
P_o = static pressure of the medium—essentially the atmospheric pressure in the absence of a sound wave travelling through the gas ($10^5 N/m^2$)
d = density of gaseous medium (kg/m^3)

Thus, the sound velocity in a gas which weighs nine times as much as air per unit volume, like Freon 12, is one-third of that in air, assuming that there is no change in the ratio of the specific heats and no change in the atmospheric pressure of the gas. In the case of Freon 12, the sound velocity has been measured to be close to one-third of that in air, so that the reverberation time of a chamber filled with dichlorodifluoromethane is three times as long as when the same chamber contains air. An even heavier gas is sulfur hexafluoride.

Similarly, for the same reverberation period, a chamber filled with Freon can be much smaller than one filled with air when both exhibit the same average boundary absorptivity a. This reduction in room volume is represented by the factor $4V/S$, which is the mean free path of the sound reflection r. In the case of a cubical enclosure with side x, when $4x^3/6x^2 = r/3$, it turns out that the Freon-filled cube may have 1/27th the volume of the larger air-filled cube for the same reverberation time.

For the general case, it can be written as

$$T_1 = \frac{13.8 \times 4V_1}{caS_1}$$

$$T_2 = \frac{13.8 \times 4V_2}{caS_2/3}$$

$$\frac{T_1}{T_2} = \frac{V_1 S_2}{3S_1 V_2}$$

$$= 1$$

$$\frac{V_1}{S_1} = \frac{3V_2}{S_2}$$

For a cube,

$$\frac{x_1^3}{6x_1^2} = \frac{3x_2^3}{6x_2^2}$$

$$x_1 = 3x_2$$

Thus, when $x_1 = 30$ ft (9.1 m), $x_2 = 10$ ft (3 m).

$$\frac{V_1}{V_2} = \frac{30^3}{10^3} = \frac{27{,}000}{1000}$$

$$\text{or} \quad \frac{V_2}{V_1} = \frac{1}{27}$$

For other parallelepiped enclosures, the ratio is similarly large. This is not to say that miniature reverberation rooms should be parallepipeds; in fact, they should not be of such a shape. They should preferably have non-parallel walls, so as to avoid coincident reinforcement of the normal modes of vibrations in the chamber and thus avoid a tone bias.

The three questions most often asked of this writer in regard to his patented device were "How well does it work," "Is such a chamber commercially available," and "Would you recommend that one build such a unit as a commercial item?"

Below is a brief history of the test data of the device, which is illustrated in Figs. 18.2 and 18.3 taken from the patent paper.

The 3 ft × 4 ft × 5 ft (0.91 × 1.2 × 1.5 m) model chamber was built of 1/16 in. (0.16 cm) thick steel, reinforced internally by bars 10 and 14 with the intent of reducing pronounced low-frequency panel resonances, and placed inside a plywood box.

Unfortunately, the model chamber walls were insufficiently treated on the outside with a viscoelastic compound. This lack of mechanical resistance on the part of the walls, plus the fact that the chamber had parallel walls, resulted in a pronounced resonance of the chamber at 407 Hz. The reverberation times for most of the audio frequency range from 100-4000 Hz, however, were well above 3 s for the 60 ft^3 (1.7 m^3) chamber, which theoretically was equivalent to approximately 1600 ft^3 (45 m^3).

Since this writer retired from RCA in 1965 he occasionally experimented with such a unit and found that the best materials for such a chamber were either lead or concrete. The concrete chamber should be painted with a sealer on the inside to prevent transpiration of the Freon through the pores of the concrete. Also, the loudspeaker diaphragm should be of plastic, and the moving microphone element should either be a coated crystal or a metallic diaphragm to minimize any corrosive effect which the Freon might have on these moving parts.

By itself, a lead sheet is not strong enough for the construction of a 5-ft (1.5 m) high chamber without buckling, so that it must be applied against a sheet of plywood or plasterboard. Because of its high internal resistance, such a lead-plywood chamber does not exhibit the pronounced metallic ring which the model steel chamber exhibited. A concrete chamber acts similarly without frequency accentuation.

We are here not concerned with the normal modes in a room, whose number is dependent only on its volume. Coincidental resonances can be avoided by slanting the walls and not employing an integral relationship

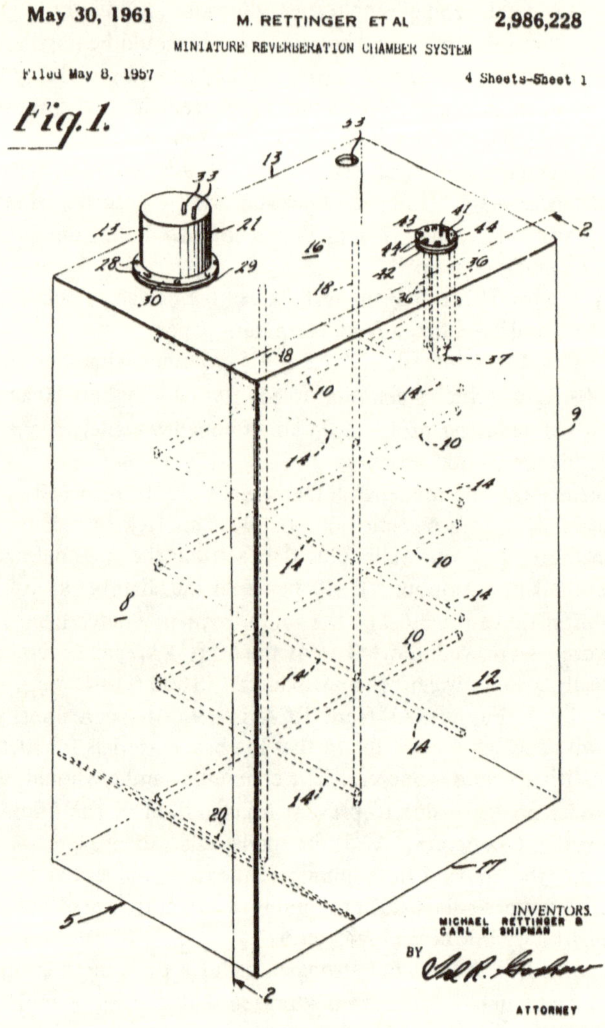

Fig. 18.2

for the three orthogonal mean dimensions.

The 407 Hz ringing of the model chamber was due to the resonances of the wall panels. Even so, the effect could be reduced considerably by the installation of a dip filter, either in the loudspeaker or the microphone line.

As far as this writer is aware, such chambers are not commercially avail-

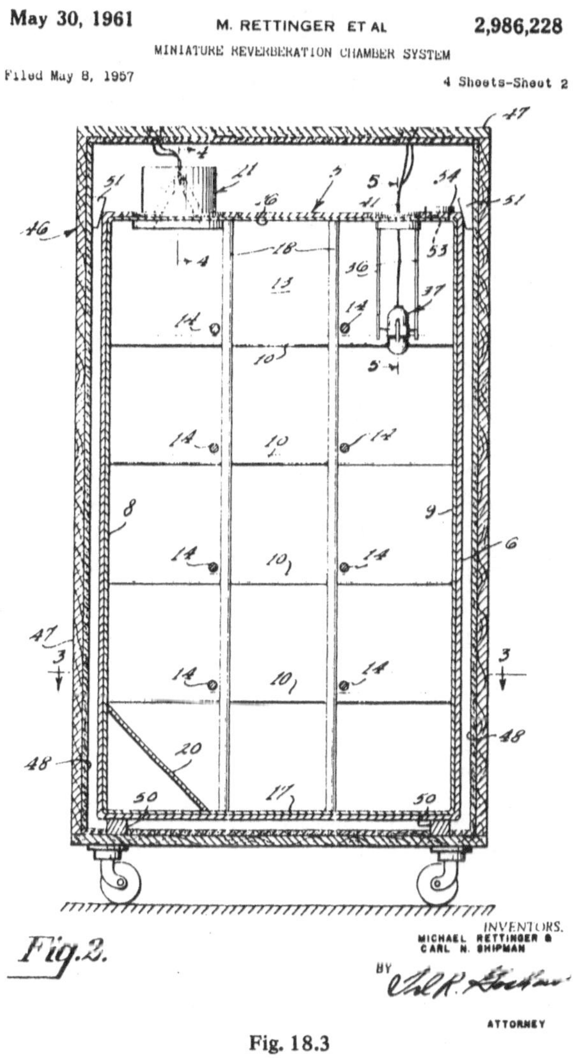

Fig. 18.3

able. RCA made a limited market survey, and felt there was not enough demand for such an item to construct it on a mass production basis. However, for someone in the recording business it would be useful to have two or three built, in place of buying EMT reverberators or other types of one-dimensional or two-dimensional sound delay devices. There is simply no substitute for using the "real McCoy," even in miniature form, when it comes to reverberation chambers with their random phase and amplitude relationships in three-dimensional reverberation.

18.4 SMALL CHAMBER

What are the acoustic conditions in a small air-filled reverberation room?

The National Bureau of Standards in Washington, D.C., built in 1976 a concrete reverberation chamber 7.4 ft (2.26 m) long, 6.24 ft (1.9 m) wide, and 5.28 ft (1.61 m) high.* The enclosure had a volume of 244 ft^3 (1.9 m^3), and was built entirely of concrete—the walls consisting of standard concrete blocks while the ceiling was cast concrete (8-in. thick). The ceiling was actually removable and was isolated from the supporting walls by a strip of Neoprene-jacketed fiberglass board. It also contained a steel door with an air-borne sound insulation rating of *STC*-50.

The measured reverberation time T at 200 Hz was 4 s and at 1000 Hz it was 2 s. From the Sabine reverberation time equation

$$T = \frac{0.05V}{aS} = \frac{0.163V'}{aS'}$$

where

S = total interior voundary
= 236 ft^2
S' = total interior boundary (m^2)
V = volume of chamber
= 244 ft^3
V' = volume of chamber (m^3)
a = average absorptivity

It can be seen that $a = 0.13$ at 200 Hz and 0.26 at 1000 Hz. This means that the total absorption A at 200 Hz was $236 \times 0.13 = 3.07$ sabins (0.28 m^2) and at 1000 it was 6.14 sabins (0.57 m^2). It could have been that at 1000 Hz the 2.7 ft^2 (0.25 m^2) of Neoprene-jacketed material at the ceiling contributed substantially to the absorption at that frequency, but this is only an assumption.

* "A Small-Scale Multi-Purpose Reverberation Room," by D.S. Pallet, E.T. Pirce, and D.D. Toth, APPLIED ACOUSTICS, October 1976, p. 287.

Excepting for the steep rise at the base, the reverberation characteristic of this small-scale reverberation chamber is remarkably smooth.

To flatten the reverberation time characteristic of the enclosure, the Bureau of Standards introduced some low-frequency absorbers in the form of thin panels, and also made sure that the relative humidity in the chamber was at least 50 percent, since a lower relative humidity tends to absorb the higher notes more than the lower ones.

The reverberation times of the small reverberation chamber before and after the introduction of some low-frequency absorbers were as follows:

	Reverberation Time (s)			
	200 Hz	500 Hz	1000 Hz	5000 Hz
Before treatment	5.0	3.2	2.0	1.8
After treatment	1.2	0.6	1.0	1.0

The question whether a reverberation period of 1 s or slightly less between 100 and 5000 Hz is sufficiently long for introducing an effective reverberatory note into a recording depends, of course, on the type of music which is to be modified, or the type of signal, like speech, which is to sound as if it were uttered in a deep cave, etc. Generally, the level of a reverberated signal may be markedly low compared to the direct sound or music level which is to be affected. Indeed, the reverberated signal may be 20 dB below the direct signal and still, the modified signal sounds remarkably reverberant. This is probably for the reason that one judges the reverberatory character of such signals not by their total decay time, but by the first 10 decibels after their decay has started. This reverberatory effect may be increased by employing a great distance between the loudspeaker and microphone in the reverberation chamber, and by placing the loudspeaker in a corner of the enclosure.

It may be noted that this writer had to install low-frequency absorbers also in a "standard-size" reverberation chamber of 2500 ft^3 (71 m^3) or larger to flatten the reverberation time characterisc of the room.

To avoid the allocation of expensive space for reverberation chambers in a recording studio, a number of artificial or synthetic reverberation generators have been employed. Among these are digital processors, steel plates, and springs.

In digital reverberation devices the amplified microphone signal voltage is sampled at a high rate and converted into digital numbers. A processor places these numbers into a memory bank and later withdraws them while simultaneously adding them or mixing them together. The mix can then be sent back to the memory bank or reconverted into the desired analog voltage, similar to the analog voltage which comes from the microphone in a reverberation chamber. Some digital reverberators have shown coloration and a rough sound decay. Coloration is the unnatural emphasis of certain frequencies in a voice or musical rendition, which in ditial processors is often rendered less noticeable by an electronic "smoothing" process during the mixing period.

A plate reverberator is essentially a two dimensional device for prolonging a signal. The two best known units of this type are the Swiss-built EMT's and the James Cunningham "Echoplate." Both are capable of prolonging the high-frequency reverberation, compared to that coming from a reverberation chamber in which it is limited by the molecular sound absorption at the treble. However, plate reverberators also tend to exhibit some coloration.

A spring reverberator is a one-dimensional unit, and possibly the least expensive of the three synthetic devices. It exhibits the most coloration and has the roughest sound decay. It also produces a "periodicity" in the reverberated signal generated by the sound traveling back and forth along the spring.

Chapter 19

Motion Picture and TV Stages

19.1 GENERAL

Motion picture and TV stages vary widely in volume and dimensional ratios. Those built before 1960 rarely exceeded a height of 45 ft (13.7 m), while later ones had heights as great as 55 ft (16.7 m), as in the case of the Hollywood ABC Stages 57 and 59 and the Osmond Studios in Orem, Utah. Not only do higher buildings provide better "bird's eye" photography but they also allow the introduction of the large cyclorama screens necessary to achieve realistic scenery inside the enclosure for medium-long shots.

Fig. 19.1 shows the dimensions of the 6 largest motion picture and TV stages in Europe and the United States, exclusive of those in existence in Russia and the People's Republic of China.

Stages are often built in twos and threes to facilitate production. Fig. 19.2 shows the plan of the American Air Lines Studios in Dallas, Texas, and Fig. 19.3 shows the plan of the Los Angeles Community College Studios for both of which this investigator provided the acoustic design. Similar plan layouts exist at the Pinewood Studios in England and at the new Sam Goldwyn Studios in Hollywood. In the latter case, the partition between the two adjoining stages can be removed in pieces by a large crane capable of lifting its concrete panels out of their recesses.

There are no equations determining the precise dimensional proportions of a TV studio, only guidelines or rules of thumb. For example, stages with heights less than 15 ft are difficult to light because of the nearly horizontal angles which the luminaire axes have to assume so that the fixtures are not visible in the scene. Such lighting provokes complaints from the performers that it is too bright and hurts their eyes. On the other hand, when the studio is much higher than 20 ft it becomes difficult to manipulate the fixtures continually and focus them for best picture quality. This is readily understood when one considers the conventional rigging from

Stage	Floor size ft	Floor size m	Floor area ft²	Floor area m²
007, Pinewood, England	160 × 374	48.7 × 114.0	59,840	5559
15, Cinecitta, Italy	136 × 311	41.4 × 94.7	42,296	3929
5, Cinecitta, Italy	118 × 262	35.9 × 72.1	30,798	2861
H, Shepperton, England	118 × 261	35.9 × 79.5	29,750	2768
Astoria, New York	120 × 217	36.5 × 66.0	26,040	2419
Osmond, Orem, Utah	110 × 150	33.5 × 45.7	16,300	1514

Stage	*Height* ft	*Height* m	*Volume* ft³	*Volume* m³
007, Pinewood, England	53	16	2,393,600	67,739
15, Cinecitta, Italy	40	12	1,691,840	47,879
5, Cinecitta, Italy	45	14	1,385,910	39,221
H, Shepperton, England	45	14	1,338,750	37,887
Astoria, New York	40	12	1,041,600	29,447
Osmond, Orem, Utah	55	17	907,500	25,682

Fig. 19.1 Dimensions of the six largest motion picture and TV stages in the world excluding those in existence in Russia and the People's Republic of China

which the lights are suspended. This consists of a grid of pipes or steel channels, usually arranged in 4-ft (1.22 m) squares, which is fixed in height. In larger studios this grid may be moved up and down, and has counter-balanced weighting along the walls of the studio to facilitate elevation changes. Note that lighting requirements are to some extent dependent on the number of cameras in use at any one time and the preferred camera zoom lenses. With more than 2 cameras in action, production time must be carefully harnessed. "Turn-around" time is the time required to strike one set and install another. This time necessarily inclues the new lighting set-up, so that the time with which the lighting can be rearranged must be kept to a minimum, particularly when the set change consists of merely moving one set out and another, prefinished one, onto the stage. In this connection it may be noted that all lighting must conform to the standard picture aspect ratio of 3 × 4, meaning that the height consists of 3 units of length and the width of 4 units of length.

Before 1950, most video control rooms could overlook the adjoining

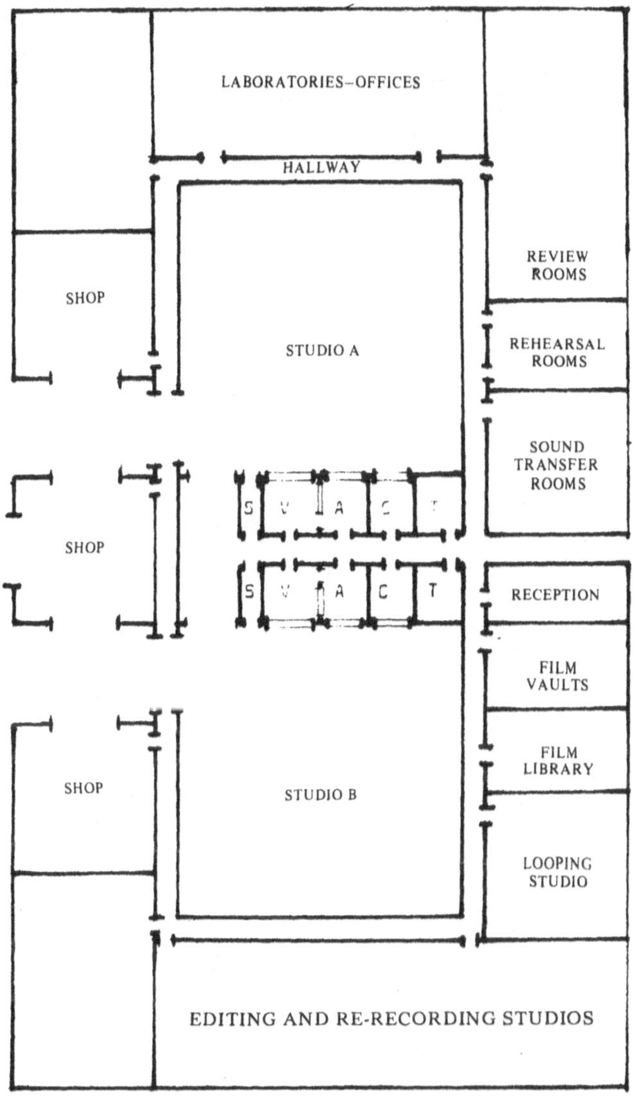

Nomenclature: V: Video Room; A: Audio Room; T: Toilets; C: Client Room; S: Sound-Lock; P: Passageways for scenery

Fig. 19.2 Proposed studios for American Air Lines in Dallas, Texas.

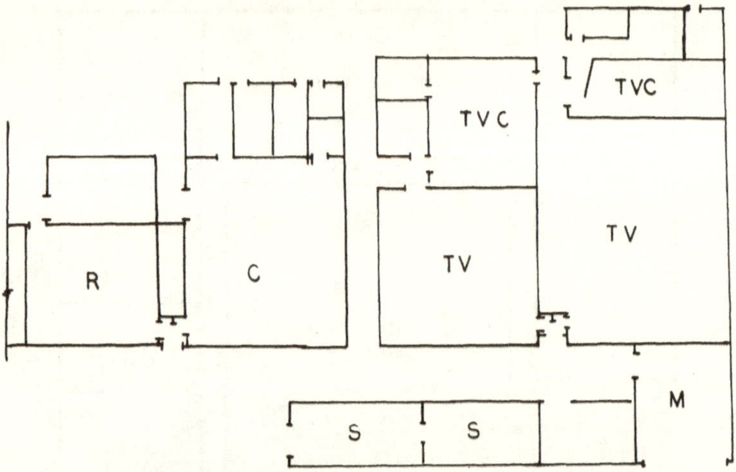

Fig. 19.3 Los Angeles Community College Studios. TV; TV stage; TVC: TV control room; R: radio studio; S: shop; M: mobile unit

stage through a window. After that period, when TV production became more complex with numerous picture monitors in the control room to permit rapid intercutting of various picture details, both the video and audio control rooms were made "blind," without a view over the stage. On the other hand, the audience-participation stage of the Osmond Studios built in Orem, Utah, in 1977, employed three in-line control rooms—two video and one audio, with the center video enclosure having a view over the stage and its audience. The mixer in the main central control enclosure could look to his right into the audio control booth and to his left into the color-control chamber. All observations were through double-pane sound-retardant windows, and doors in the two partitions allowed ready travel from one room to another.

In some instances a video truck storage space is desired near a TV stage to permit ready interfacing between the equipment in the truck and that in the control rooms.

In the case of audience-participation stages, moveable bleachers are recommended so that, without an audience, the entire stage floor is available for production. In the case of the Osmond Studios it was also possible to slide open a 20 ft × 45 ft (6.1 m × 13.7 m) sound-retardant door to allow the rolling in of an ice rink during a program when a dancing performance had to be almost immediately followed by a skating performance.

Cyclorama screens (many seamless) have become essential elements in

TV studios. Acoustically they are low-frequency absorbers with a resonance frequency given by (see Fig. 19.4)

$$f_0 = \frac{60}{(MD)^{1/2}}$$

or

$$f_0 = \frac{170}{(md)^{1/2}}$$

where

M = surface density of screen (kg/m²)
D = distance between screen and hard wall in back (m)
m = surface density of screen (lb/ft²)
d = distance between screen and hard wall in back (in.)

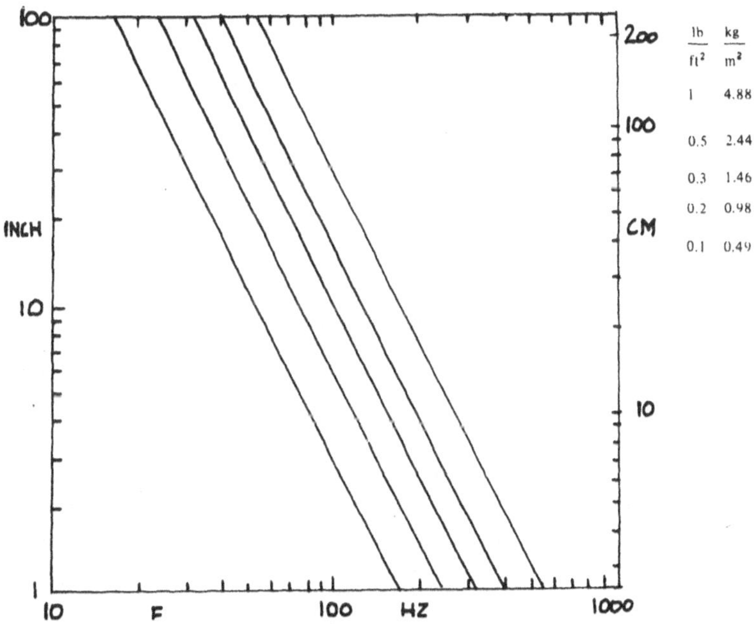

Fig. 19.4 Variation of panel resonance with distance between panel and hard backing, for various surface densities of panel

When the wall in back of the screen is hard, and the distance between the screen and the wall is a half wavelength of the incident sound, minimum absorption occurs at the corresponding frequency and even multiples thereof; when the distance represents a quarter wavelength, maximum absorption occurs at the corresponding frequency and odd multiples thereof. This is illustrated in Fig. 19.4.

To smooth out the peaks and dips of the absorption characteristic of the cyclorama screen, the wall in back of the screen should be treated sound-absorbently, as shown in the above figure.

When the studio has insufficient height for uniformly illuminating the cyclorama from the top, a 3 ft × 3 ft (0.9 m × 0.9 m) groove or channel may have to be provided in the floor of the studio, into which the screen can be sunk. This allows the illumination of the screen from the bottom also without showing the luminaires in the channel.

Other TV stages have wrap-around cyclorama screens with radiused or concave white coves at the top and bottom to minimize an abrupt change in the illumination at floor level and ceiling height.

A type of portable cinema or TV stage consists of a large plastic hemispherical or rectangular balloon into which air is forced. An "air bubble" of this type can be inflated in less than a day with a blower of sufficient capacity. For instance, a hemispherical balloon with a circular base area of 10,000 ft^2 (929 m^2) has a volume of 378,000 ft^3 (10,697 m^3). A blower with an air flow capacity of 378,000/24 = 15,750 ft^3 (446 m^3) per h or greater can inflate the bubble in 24 h.

Such a structure is not airtight and the blower has to be continually in operation, although it may do so at a reduced speed. When warm air is pumped in, less continuous blower operation is necessary, because of the buoyancy effect of the lighter warm air.

During dialogue recording the blower is generally shut off to avoid noise pick-up. Nonoperation of the blower may be for more than an hour before serious sagging of the walls is noticed.

The plastic may be light-transparent, translucent, or somewhat opaque. It should be nonflammable and water resistant.

Adequate "air lock" entrances are required to avoid excessive blower operation, and the blower may have to be put in a shed outside the balloon should rainy weather appear.

Obviously the sets can not very well be supported from the dome and must be mounted entirely on the wood floor if one is built for the unit, or the ground itself.

Units of this type have also been used for temporary warehouses, geologists' shelters on the desert or on an island, and even for exhibit facilities, like the U.S. Pavillion at the 1970 Exposition in Japan.

Another type of portable sound stage consists of a double-skin air bag, either in the shape of a hemisphere or half a cylinder. The lower portions of the structure are anchored to the ground, and an air compressor forces air into the bag. It can be inflated in a few hours, and is completely waterproof so that it may even be used during a rain. One such device is made by Air Tech.*

19.2 STAGE FLOORS

When a concrete floor slab rests on a layer of sand or decomposed granite, the base acts as a cushion for the slab. The relatively soft bed absorbs a great deal of the vibrational energy when the source is on the floor and, similarly, protects the concrete floor and its sensitive load against the infiltration of external vibrations in the ground.

It is difficult to assign elastic constants to a granulated base because of the almost infinite variety of such unconsolidated heterogeneous material of which much of the earth's crust is composed. Hence a range of values should be used in the calculations of the resonance frequency of the cushioned slab unless exact values for the dynamic modulus of elasticity of the weathered rock, its Poisson's ratio, density, etc., can be established by tests.

For a concrete floor slab with an area larger than 1000 ft^2 (93 m^2) and a thickness of 6 in. (15 cm), the resonance frequency is close to 26 Hz for a sand thickness of 6 in. (15 cm). For a concrete floor thickness of 12 in. (30 cm), the frequency would be $26/(2)^{1/2}$ or approximately 18 Hz, because the static compression of the sand would be twice as large (Hook's law remaining active). This is because the resonance frequency is inversely proportional to the square root of the static deflection.

For concrete floor areas smaller than the indicated range of 1000 to 10,000 ft^2 (93 m^2 to 930 m^2). The resonance frequency becomes higher, because the static compression of the sand increases inversely with the area that is loaded. Thus, for a concrete floor of 500 ft^2 (46 m^2), the resonance frequency could be $(2)^{1/2}$ as high, and one of 250 ft^2 (23 m^2) could be twice as high.

---

* located at 85 Madison Circle, East Rutherford, N.J.

To reduce the transmission of vibration into the stage, the concrete floor should be cast on a layer of 6 in. (15 cm) thick sand or decomposed granite, with care being taken to prevent the concrete fluid from running into the sand and solidifying it (see Fig. 19.5). The perimeter of the floor

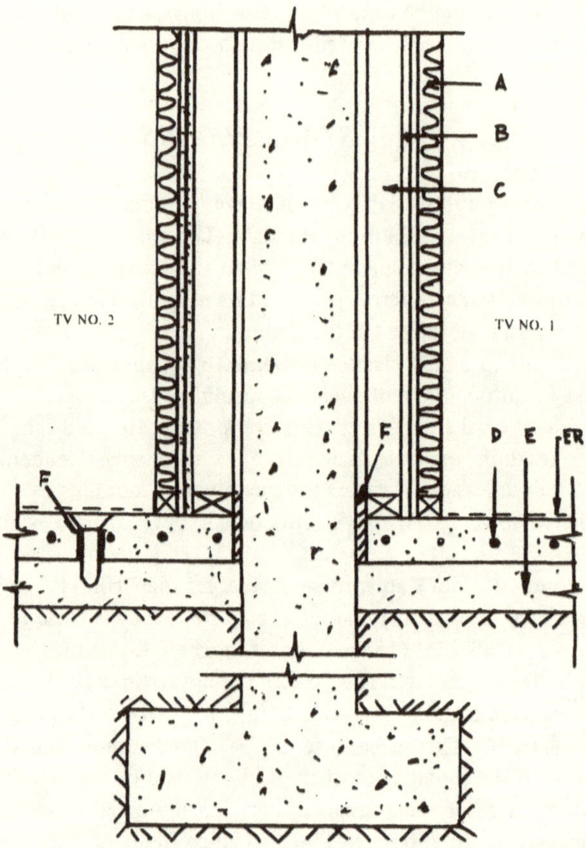

Fig. 19.5 Cross section through two adjoining TV stages. A: acoustic treatment; B: 2 sheets of ½ in. plasterboard; C: steel stud; D: reinforcing bar; E: 4 in. sand; F: plastic sealer; ER: 3/16 in. epoxy resin.

should not touch the foundation but should be isolated from it by 2 in. (5 cm) thick asphalt-impregnated fiberglass board. Because the ratio of the density of concrete to that of sand is close to 1.4, and the ratio of the longitudinal velocity of sound in concrete to that in sand is close to 20, the impedance mismatch of the two media results in a vibration reduction of approximately $10 \log 1.4 \times 20 = 14.5$ dB over much of the audio range.

Fig. 19.6 shows a schematic view of such a stage. Note the cyclorama trough which allows the lighting of this screen from the floor (as well as from overhead) and also acts as a passage way for the air inside the stage to travel under the screen towards the wall exhaust grilles. When rain scenes are intended to be photographed on such a stage, drains should be incorporated into the floor for the rapid removal of the water. For this type of operation it is also desirable to float ¼ in. (0.63 cm) thick gray epoxy on the concrete floor to prevent moisture infiltration, and to allow the painting of this surface with water color paint, as is often desired in TV operation, without causing a permanent stain in the concrete.

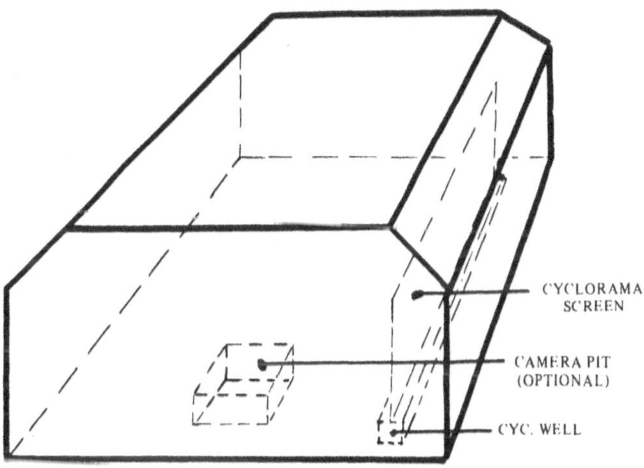

Fig. 19.6 Perspective view of TV studio.

19.3 STAGE ROOFS

Various shapes and types of construction have been employed for studio roofs. In quiet suburban areas, usually before 1930, lattice or lamella arches were chosen for this structural element because of the economy of fabrication. Later, trusses became necessary for the roof to carry the material load to exclude external noises from the stage. The table below gives the surface densities of the roof structure to achieve various sound transmission losses:

Transmission Loss	*Surface Density*					
	125 Hz		250 Hz		500 Hz	
	lb/ft^2	kg/m^2	lb/ft^2	kg/m^2	lb/ft^2	kg/m^2
38 dB	70	342	28	137	11	54
40	100	488	40	195	15	73
42	140	683	55	268	20	97
44	180	878	70	342	28	137

When the external noises are not predominantly aircraft flyover disturbances but street traffic din, which may not be able to strike the roof directly but only by diffraction at the upper edges of the building, considerable noise reduction on the roof may be obtained by building a parapet at the frontal edge to increase the barrier effect.

Roof surface densities not in excess of 12 lb/ft^2 (58 kg/m^2) can be achieved with 3 in. (7.5 cm) thick wood. For higher surface densities concrete is in order. A 1 in. (2.5 cm) thick layer of this material can generally not be poured as a single surface, because of cracking, and hence must be cast in 20 ft × 20 ft (6 m × 6 m) panels separated by ½ in. (1.25 cm) thick expansion joints. Lead sheets are generally uneconomical for a roof load on account of the high price of lead (close to a dollar per pound in 1979).

19.4 WALLS

Because pound for pound, concrete is the cheapest building material, many TV and cinema stages about urban areas employ it in their wall construction. In Hollywood, numerous stages use the so-called tilt-up method of wall fabrication, whereby a slab of concrete is poured on the ground and then raised into a vertical position with a crane. It is a fast system of building, but requires considerable experience on the part of the designer and contractor.

Presently, (as of 1980) concrete block walls dominate the scene. They may be single tiers or so-called cavity walls, which consist of two tiers or wythes of masonry separated by a continuous airspace not less than 2 in. (5 cm) wide. The exterior or facing tier is always a 4 in. (10 cm) unit, while the interior one may be 4 in., 6 in., or 8 in. (10 cm, 15 cm, 20 cm), depending on the structural requirements of the wall.

The sound insulation of a cavity wall can be increased considerably by omitting the metal ties between the wythes, whereby a double-stud wall will also have a 6 to 10 greater noise reduction than a single-stud wall of the same surface density. When lateral support is called for between the two walls, a resilient one should be used, as illustrated in Fig. 10.5. Such a unit has a load-bearing capacity of 100 lb (45 kg) both in tension and compression. The sound insulation may be made still greater by increasing the airspace to 4 in. (10 cm) and by inserting a 2in. (5 cm) thick fiberglass board to reduce the sound pressure in the interspace and to damp any standing waves which may form therein. Note also that painting or plastering the concrete block wall is desirable to reduce its sound transmission capability, as shown in Fig. 19.7.

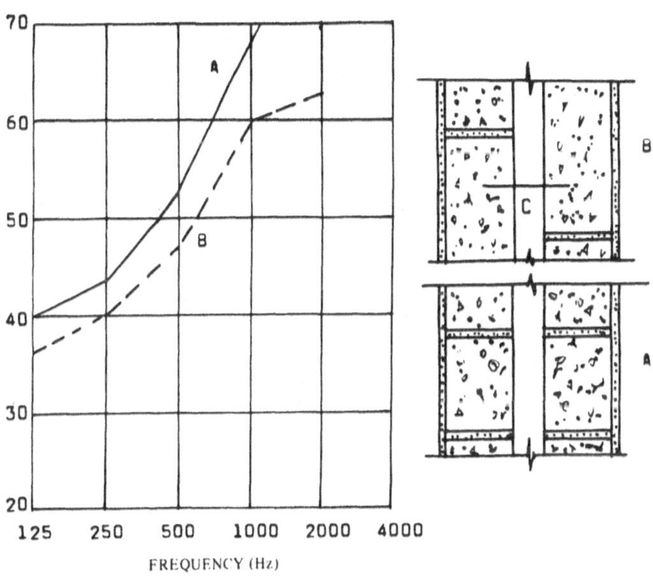

Fig. 19.7 Sound transmission loss characteristic of concrete block cavity wall with (B) and without (A) metal rod ties (C)

Chapter 20

Music Recording Studios

Since a microphone represents one ear, its limitations should be recognized by the operator of a studio. For instance, the reverberation of the studio when reproduced in the control room appears longer to a listener there than it does to one in the studio. Similarly noise in the studio appears louder in the control room than in the studio. These and other effects occur because the microphone is a monaural listening device. An observer listening with two ears is able to suppress unwanted signals like noise and reverberation, which is a peculiarly subjective quality in humans. When an observer plugs one of his or her ears, the room in which the observer is situated will sound noisier and more reverberant to this observer.

In the design of a sound recording studio, therefore, it becomes necessary to effect a low noise level therein as well as an adequately short reverberation time.

Another important factor in such a room is its interior decor. Musicians consider themselves artists and are inspired by the attractive appearance of the space in which they perform. In the forefront of the desirable materials for such an enclosure is wood, preferably one with a distinct grain and a smooth polished surface. Musicians have learned long ago that violins, violas, violin-cellos, and others, when made of aluminum or plastic, sound inferior in tone quality to such instruments built of wood. To this extent, the wood on the walls and floor are by many considered to be extensions of the instruments.

Needless to say, a studio should be adequately air-conditioned to allow sustained efforts by the performers without tiring them.

Rarely is a music-recording studio used exclusively with one group of instruments, one type of music, and one system of sound recording with a set number of microphones. More than likely, the number of musicians will vary from day to day, as may the type of music (rock-and-roll, classical, religious, etc.), and the number of sound tracks later employed for "mix-down." In some studios, photography may take place at times, and in all, the music of "takes" or other recorded renditions is likely to be reproduced in various manners, including monophonic, stereophonic, triphonic, quadraphonic, quintaphonic (as in the Columbia Pictures production "Tommy"), and other systems. In some enclosures, 16 mm or 35 mm motion pictures may be projected onto a screen for "post-scoring" or video-disk reproductions may be seen on a TV receiver screen during recording sessions. In short, sound-recording studios are like people, in that no two are alike.

When then, the operating conditions of a music-recording studio tend to vary, what design philosophy should be followed? The answer must lie in the introduction of acoustic variability. The room should exhibit control of reverberation as well as adjustment of the localized liveness about a performer or group of performers. It may also feature instrument cells for the placement of band sections in multiple track recording, associated vocal rooms, chorus booths, etc.

Acoustic variability in a room may be achieved in several ways. In one instance it was possible to change the volume of the studio by moving one of the sidewalls; in another, by lowering the ceiling. In the majority of studios, however, recourse is taken by changing the absorptivity of one or more of the walls.

Figs. 20.1 and 20.2 show details of the construction of rather elaborate vertically rotatable, double-hinged splays which are absorbent on one side and reflective on the other. By orienting the splays properly, one or more walls of the enclosure may be made entirely absorbent or reflective, or partially so. Note that when a wall is to be completely reflective, it is made so by adjusting the splays to form triangular corrugations. Thus, when the opposite wall is also reflective, no flutter echoes can occur in the manner in which they would occur when the reflective sidewalls are flat and parallel. This is accomplished by using "low" and "high" splays. While all splays are exactly alike, they do differ in their wall hinges, one having a height of 2.5 in. (6.4 cm) and the other one of 1 in. (2.54 cm).

The vertically rotatable double-hinged splays can function also as instrument cells for multiple-track recording, so that the microphone for each instrument or instrument group picks up only the sound from it and not also the sound from an adjoining player or players. Some music directors prefer a small plastic window in each splay so that adjoining musicians can see each other and can thus play in better time and tune.

Splays higher than 8 ft (2.4 m) should have a caster under their outer edge for better weight support—the splays being estimated to weigh 2 lb/ft^2

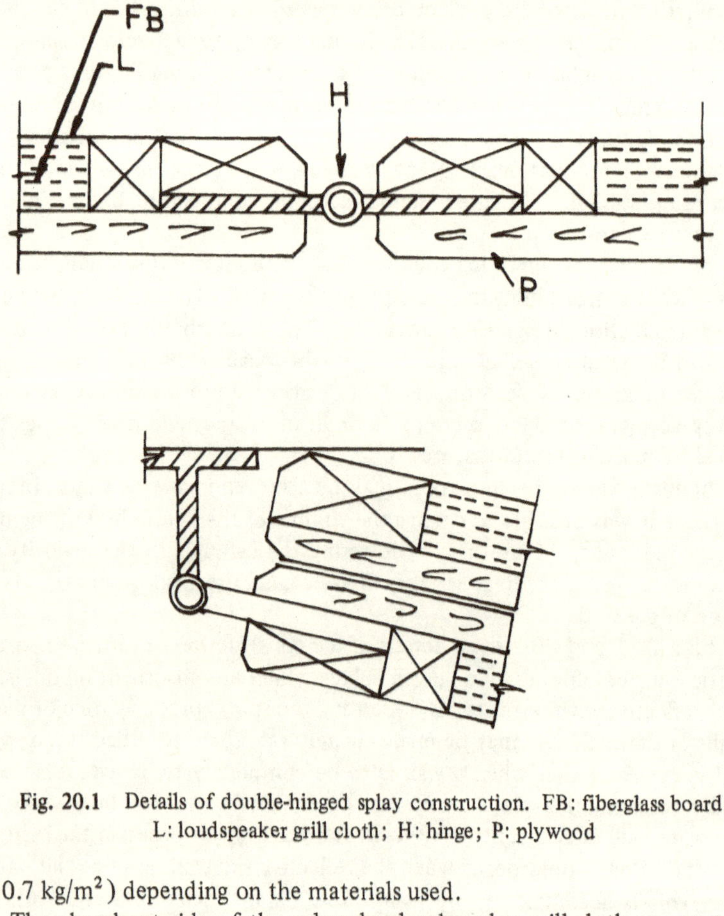

Fig. 20.1 Details of double-hinged splay construction. FB: fiberglass board; L: loudspeaker grill cloth; H: hinge; P: plywood

(10.7 kg/m^2) depending on the materials used.

The absorbent side of the splays has loudspeaker grill cloth as a protective surface for the fiberglass board. So-called "pegboard" or 1/8 in. (0.317 cm) thick perforated Masonite is not recommended as a protective cover, because such surfaces give rise to high-frequency echoes. Decorative vinyl may also be employed, but it should be sound-trasparent by employing the highly perforated kind. It is also desirable to fit the splays with retractable handles for their easy turning and extensions.

When the reversible splays are intended as instrument isolation cells, their lower edges should be kept as near to the floor as possible to prevent sound transmission under them from one cell to another. If that is not

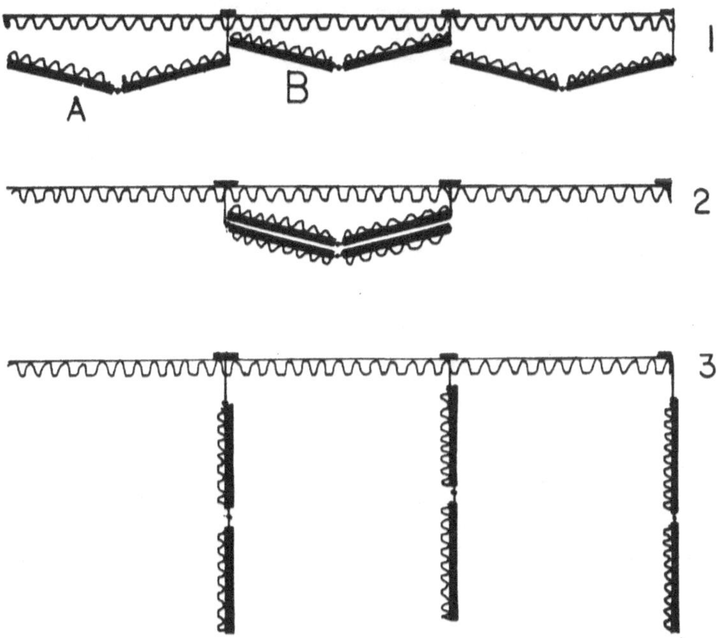

Fig. 20.2 Diagram 1: Double-hinged high (A) and low (B) splays closed, exposing solid (reflective) surfaces to studio interior. Diagram 2: High double-hinged splays folded over low ones, making the wall entirely absorbent. Diagram 3: Splays are open to act as instrument cells for multiple-track recording.

readily possible, as when casters have to be put under the splays to help support their weight, a drop-bottom felt pad should be attached to the side of the splay near its lower edge, so that it may be lowered when the splay is in the desired position and the space between the lower edge of splay and floor is closed to stop this sound transmission. Of course, the height of the splays should be as great as possible to avoid sound transmission over the upper edges from one cell to an adjacent one.

Fig. 20.3 also shows a folding partition which may be used for isolating instruments. The partition extends like an accordion, and may be pushed into a wall receptacle when not in use.

The use of singly hinged instead of doubly hinged vertically rotatable panels, hard on one side and soft on the other, is less expensive, and many

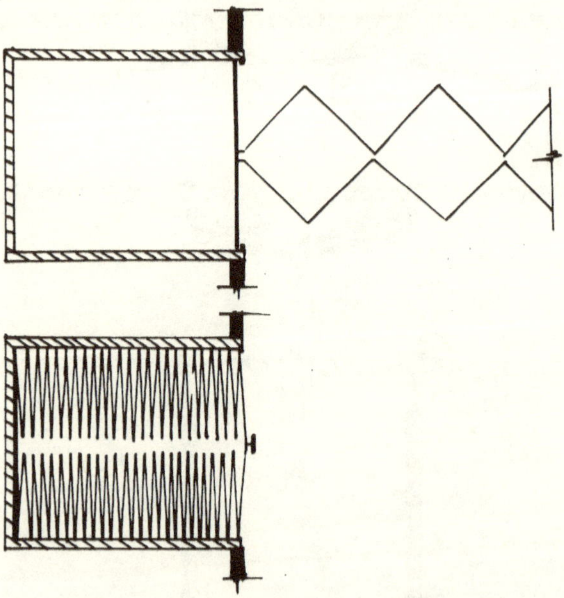

Fig. 20.3 Folding-type space divider

studios have employed them successfully.

Some studios have enclosures for drums and piano built into the side-walls. In the case of piano booths it is not always necessary to isolate the instrument to a high degree, since its music may "tie" the various tonal sources on the stage together, so that a glass sliding door is often installed for that booth, as shown in Fig. 20.4. Also, closed-circuit television may be employed to allow the instrumentalists to see each other. The glass sliding door may also serve to adjust the liveness of the enclosure.

So-called drum cages are enclosures in which drums may be isolated from the other instruments of a band. Such rooms often carry little sound-absorbent materials, because drummers insist that their music does not sound pleasant to them in dead cubicles. For this reason drum sounds are sometimes not recorded with microphones but with accelerometers attached to the instruments. Such recordings sound surprisingly "clean"—sharp staccato pulses with little of the reinforcement produced by the room. Of course, the recording engineer may then add any amount of

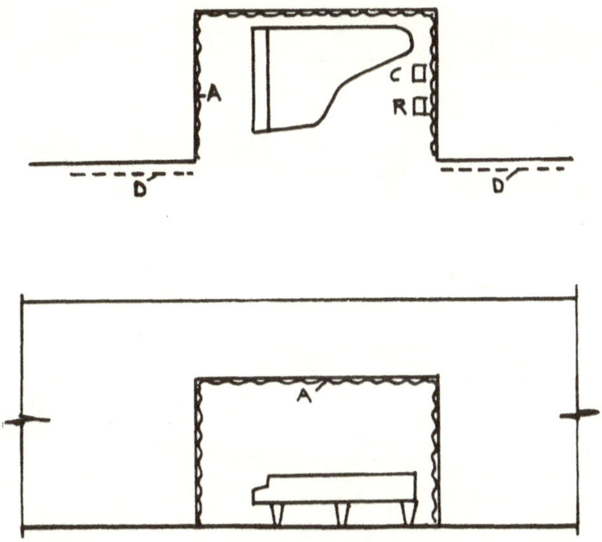

Fig. 20.4 Piano booth. A: absorbent treatment; D: glass sliding door (optional); C and R: camera and receiver for closed-circuit television.

electronic reverberation to the recording in order to improve the recording sound.

Figs. 20.5 and 20.6 show photographs of a large music recording studio designed by the author. Note the cylindrical splays on the walls and ceiling which run at right angles to each other for increased sound diffusion in the studio.

Rooms in which sound recording takes place or in which critical listening is to be done, as in a monitoring enclosure, the floor and ceiling should preferably not be parallel. This is true even in the case of a so called suspended acoustic ceiling, because this ceiling with its constant depth airspace behind it exhibits sharp dips and peaks in its absorption characteristic. Such a ceiling should slant slightly, either over its entire length, or in sections from 4 to 10 ft in length, depending on the volume of the enclosure.

A suspended acoustic ceiling in a "critical" room is also desirable when the floor is covered with a carpet. This is because such a material is deficient in low-frequency absorption, even though at 1000 Hz a ¼ in. (0.63

Fig. 20.5 Republic Studios scoring stage, view towards projection and control rooms.

cm) or 3/8 in. (0.95 cm) thick carpet may show an absorption coefficient as high as 50 percent, particularly when an Ozite of foam rubber underlay is employed below the carpet. The suspended acoustic ceiling with its relatively high low-frequency absorptivity, on account of the airspace, tends to compensate for the deficiency in the low-frequency absorption of the carpet.

A suspended ceiling is also able to hide air-conditioning system ducts and allow recessed lighting.

When the floor is highly absorbent with a thick carpet and thick pad, it is sometimes desirable to intersperse the acoustic lay-in panels on the ceiling with hardwood plywood panels or plasterboard, so as not to create too dead an airspace for the performers in the room.

Triangular and cylindrical wall and ceiling splays were in great vogue before the days of multiple track recording, when microphone distances were rarely less than 6 ft. When a studio has to be kept on the dead side to achieve "clean" tracks at every one of the several transducers, such

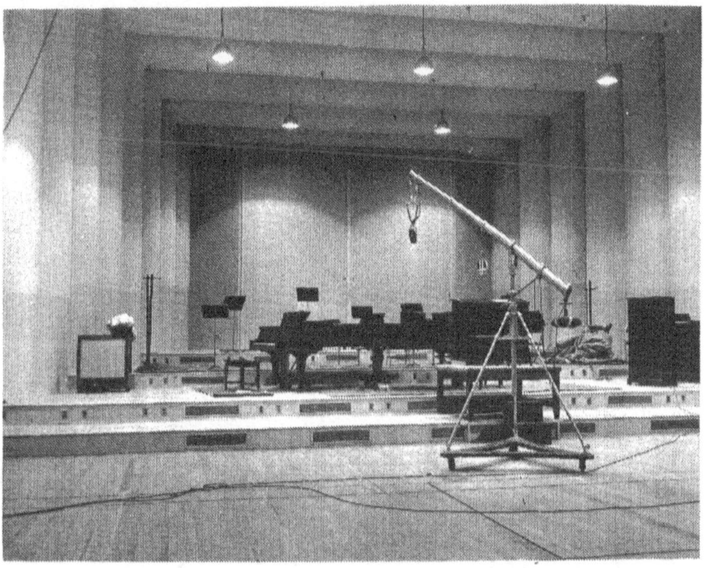

Fig. 20.6 Republic Studios scoring stage, view towards podium.

sound diffusers appear less necessary. However, they can also act as low-frequency absorbers, as shown in Fig. 20.7. Where the interior decor of a studio is highly important, the use of 1/4 in. (0.63 cm) or 3/8 in. (0.95 cm) thick hardwood plywood splays is encouraged. In some instances, 1/8 in. (0.31 cm) thick hardwood plywood is cemented to the 1/4 in. (0.63 cm) or 3/8 in. (0.95 cm) thick fir plywood forms after their installation, for economy reasons as well as to prevent cracks from showing in the wood when thicker panels are forced into curvilinear shapes (a condition which does not exist when the splays are made triangular). It is generally best to introduce sound-absorbent material on the inside of the splay to increase its low-frequency absorptivity and to prevent sharp resonance peaks. These peaks are sometimes difficult to detect by placing a microphone near the splay, because such resonances are not able to set an entire splay into vibration—only a small area of it. The reduction of such resonance peaks are readily demonstrated, however, by attaching a vibration pickup to the splay itself, before and after introduction of the sound-absorbent

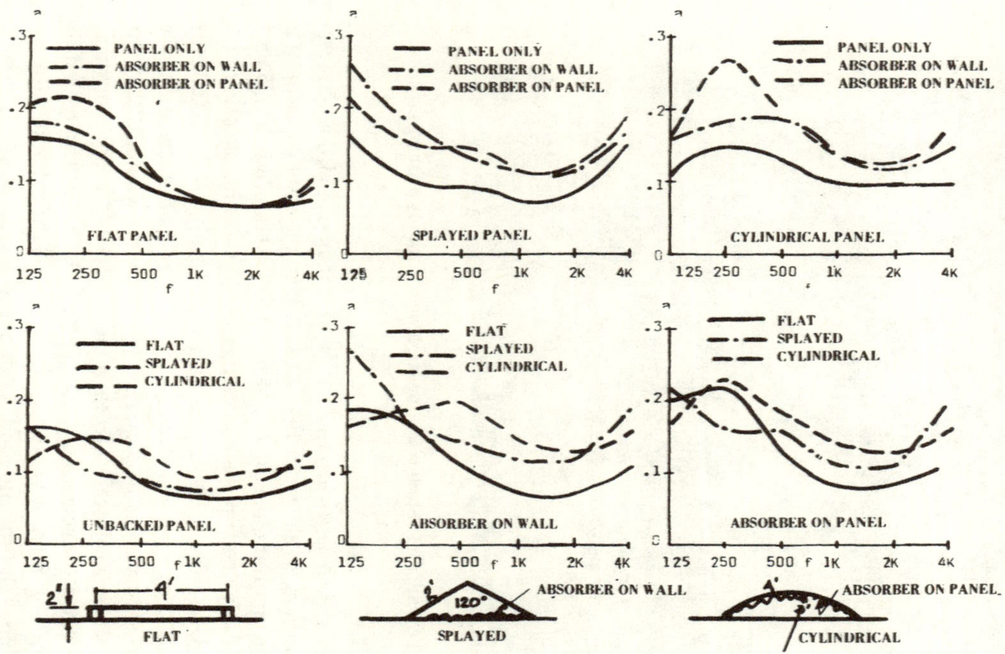

Fig. 20.7 Sound absorption characteristics of ¼ in. (0.63 cm) thick plywood panels arranged against a hard wall either as a flat, triangular, or cylindrical splay, with and without sound-absorbent treatment between the panel and the wall.

material, and exciting the splay into vibration by exposing it to a varying sinusoidal signal of high amplitude (particularly when the material is applied to the panel itself instead of on the wall within the splay). Such splays may also be made of plastic, as in the case of the huge stage splay at the Dorothy Chandler Pavillion in Los Angeles, California.

Not much has been published regarding recording studio shape and volume as a function of the number of instruments to be recorded therein. In the motion picture industry, a specific volume of 3500 ft^3 or 1000 m^3 per instrument is generally employed to arrive at the volume of the planned scoring stage. One reason why so large a space per performer is necessary for scoring stages is that they have to incorporate a wide screen for the projection of a scene for which music has to be recorded. Another reason lies in the fact that in so complicated an operation it is best, for multitrack recording, to be able to space the band sections widely apart, so that some sections of the orchestra do not have to return on another day at enormous expense.

For audio recording studios the specific volume may be one-third that of a scoring stage, unless TV cameras are used for video disk or cassette recordings. In this case, the ceiling height of the studio should not be less than 13 ft, so that the luminaires can well stay out of the video camera angle and are not able to direct too much heat onto the performers.

For studios with volumes larger than 35,000 ft^3 or 1000 m^3, a desirable ratio of dimensions is 1:3.2:3.8, to achieve maximum floor space for the musicians and hence maximum separation between instruments. The wide room also allows the construction of adequately large audio and video control rooms (for large studios two video control rooms may be necessary), plus a client's observation enclosure along the same rear wall so that important visitors can be kept out of the monitoring chambers.

Another reason for the 1:3.2:3.8 dimensional ratio is that it avoids the coincidental reinforcement of the normal modes in a room, distributing the characteristic resonance frequencies uniformly over the audio frequency range. It is based on the cube root of two or multiples thereof: $1:2 \times 2^{2/3}:3 \times 2^{1/3}$.* Another very similar ratio is 10:31:37, since 31 and 37 are primes, without a divider.

The table below gives the equations for the height, width, and length of an audio recording studio as a function of the number of performers which it is to accommodate when it is intended for video work too. The equa-

* Volkmann, J.E. "Polycylindrical Diffusers in Room Acoustic Design," Jl. Acost. Soc. Amer. V. 13, No. 3, p. 234

tions are based on a specific volume of 1000 ft^3 or 28 m^3 per instrument, as well as on a specific volume of 1500 ft^3 or 42 m^3 per instrument when the number of musicians is greater than 30; when it is smaller, the dimensions for the width and length of the studio are the same as those shown in the table, except that the ceiling height remains at 13 ft or 4 m, as shown in Fig. 20.8.

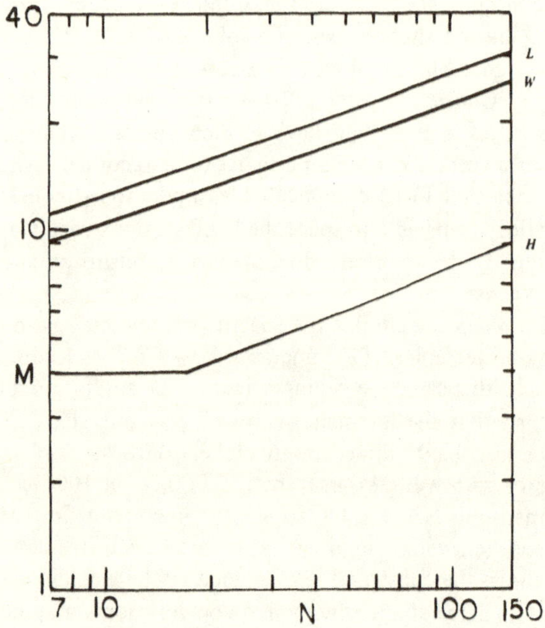

Fig. 20.8 Variation of height (H), width (W), and length (L) of music recording studio in meters with the number of instruments N, based on a specific volume of 42 m^3 per instrument.

Specific Volume	*Height*		*Width*		*Length*	
	ft	m	ft	m	ft	m
1000 ft^3 28 m^3	$4.35N^{1/3}$	$1.32N^{1/3}$	$13.9N^{1/3}$	$4.24N^{1/3}$	$16.5N^{1/3}$	$5.04N^{1/3}$
1500 ft^3 42 m^3	$4.98N^{1/3}$	$1.51N^{1/3}$	$15.94N^{1/3}$	$4.8N^{1/3}$	$18.92N^{1/3}$	$5.77N^{1/3}$

What about the shape of the studio? Undoubtedly the rectangular plan is the least expensive to build. On the other hand, a trapezoidal shape makes it easier for the mixer to see the various band sections in multiple track recording, as illustrated in Fig. 20.9.

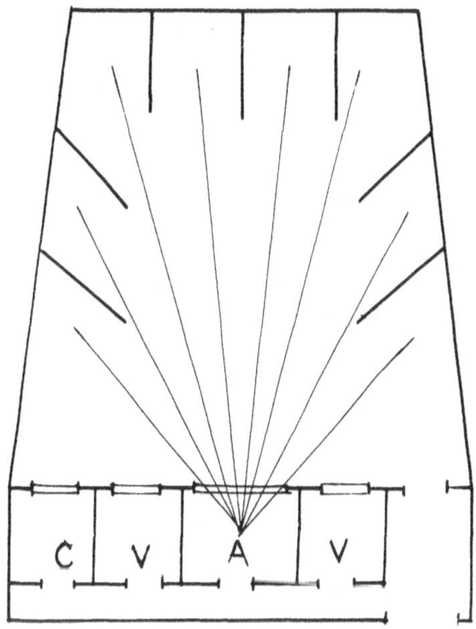

Fig. 20.9 Trapezoidal plan of recording studio allows mixer to have a view to every band section. V: video control room; A: audio control room; C: client's room.

In respect to the normal modes of vibration in a room, the existing equations to determine their spacings from each other are valid only for empty rectangular enclosures and become unreliable when even one side deviates from orthogonality, when baffles or corrugations are installed in the room, or when persons and instruments occupy considerable floor space. The total number of normal modes is a function of the room volume only, and cannot be changed by nonrectangularity, splays, etc. The selected ratio of dimensions of 1:3.2:3.8 or 1:3.1:3.7, however, does provide some insurance against colorations, that is, audible periodic reflections or strong superposed isolated reflections noticeable particularly during sound decay.

Because of the low ceilings, for which a 13-ft height corresponds to the wavelength of an 87 hertz frequency, it may be desirable to introduce ceiling splays to minimize standing wave effects unless that surface is made highly absorbent. Such splays may even add to the attractiveness of the room, and when judiciously placed may direct reflections to a desirable location. Fig. 20.10 shows several configurations of these splays. In one motion picture scoring stage the ceiling splays were made rotatable, reflective on one side and absorbent on the other, to provide a degree of acoustic variability in the room.

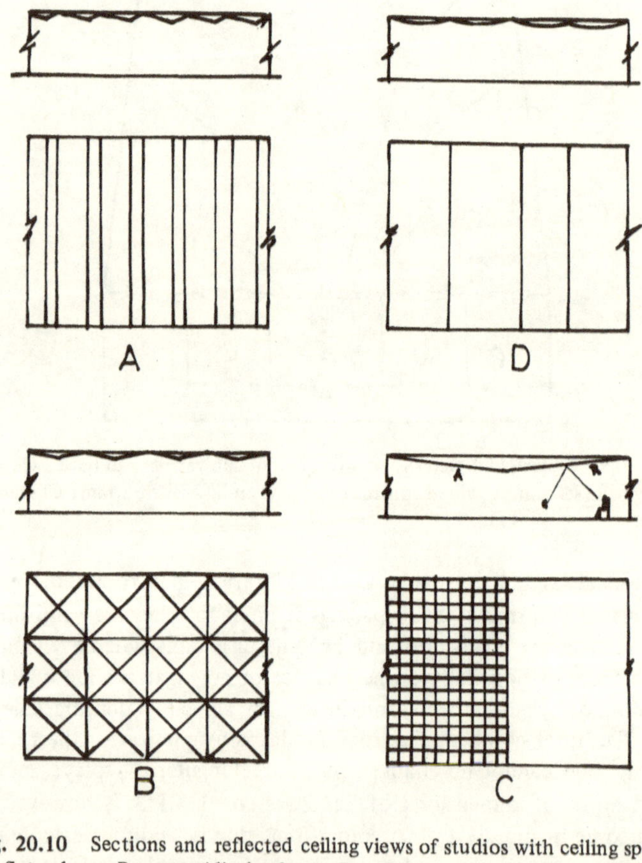

Fig. 20.10 Sections and reflected ceiling views of studios with ceiling splays. A: flat splays; B: pyramidical splays; C: reflective and absorbent splay; D: cylindrical splays.

Fig. 20.11 shows range in the 1000-Hz reverberation times found in satisfactory recording studios. This period is quite subjective, and tends to vary also with the type of music being recorded when the studio volume remains constant.

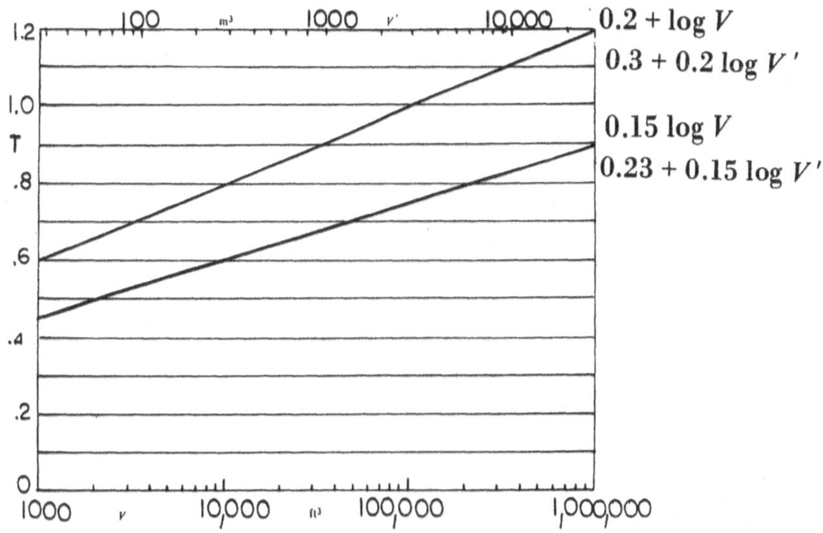

Fig. 20.11 Variation of reverberation time with recording studio volume, at 1000 Hz

In evaluating the required sound absorption in a room with a relatively short reverberation time, it is advisable to employ the so-called Eyring-Norris equation, because by the Sabine equation the absorption turns out to be larger than necessary. In the Eyring-Norris formula, however, the significant factor is the average sound absorption coefficient $\bar{a}$ for the room, which is often determined by a trial-and-error method. To facilitate this calculation, Fig. 20.12 has been prepared. It represents the equation

$$\bar{a} = 1 - \frac{1}{10^{0.05V/2.3ST}}$$

$$= 1 - \frac{1}{10^{0.163V'/TS'}}$$

where

V = room volume (ft^3)
V' = room volume (m^3)
S = interior surface (ft^2)
S' = interior surface (m^2)
T = reverberation time
$\bar{a}$ = average sound absorption coefficient

$$= \frac{\text{sum of } S_i a_i}{S} = \frac{\text{sum of } S'_i a_i}{S'}$$

S_i = surface having absorption coefficient a_i

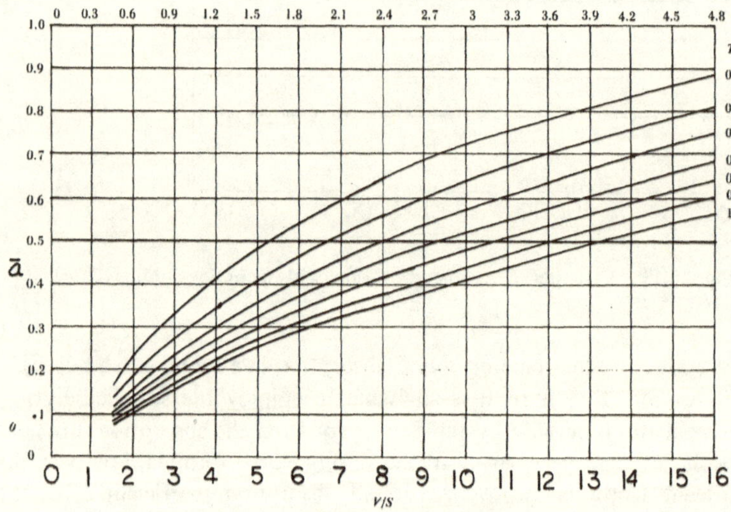

Fig. 20.12 Variation of average sound absorption coefficient with ratio of room volume V in cubic feet to interior surface S in square feet, and with ratio of room volume V' in cubic meters to interior surface S' in square meters, for different reverberation times T.

As an example consider a 13 ft × 42 ft × 49 ft studio with a volume V of 26,754 ft^3, an interior surface S of 6482 ft^2, so that V/S becomes 4.13. For a 1000-Hz reverberation period of 0.5 s, $\bar{a}$ comes to 0.34, as shown by a dot on the figure. It becomes a matter now of adjusting the individual absorption areas in such a manner that the total absorption divided by the total room surface comes to 0.34. By the Sabine equation the average absorptivity would have come to $0.05 \times 26{,}754/0.5 \times 6482 = 0.42$.

Chapter 21

Review Rooms

In some motion picture studios a review room is also called a screening room. It is intended for observing film sequences as well as entire features. It serves, therefore, as a theater in the building and should exhibit all the functions of a public motion picture theater, from variable masking of the screen to accommodate various film formats, to house light dimmers, to one or more two-way loudspeaker systems to reproduce monophonic, stereophonic, quadraphonic, etc., types of sound tracks.

To be able to judge the recorded sound on film or video tape adequately, it becomes necessary that the air-conditioning system noise be kept at the adequately low level of 30 dB-A.

There should be a volume control in the seating area of the room so that the sound level of the reproduced signals may be of satisfactory amplitude for critical sound judgements.

Similarly, noise from the projection booth should not be transmitted audibly into the review or screening room. This can be accomplished by a double-pane observation window and a single ¼ in. (0.63 cm) thick clear plate or optical glass in the projection port.

The acoustic treatment of the small viewing enclosure should be able to reduce reflections from the rear wall to a minimum. The rear wall refers to the partition between the theater proper and the projection booth. It is desirable to have some first-order sound reflections from the front parts of the sidewalls of the room and its ceiling, but second-order reflections from the rear wall tend to mask syllables and to reduce the speech intelligibility.

As noted in the section on rerecording studios, rooms with film projection should not have white but colored walls and a colored ceiling so as not to reduce the contrast on the screen.

As a further precaution against "washed-out" pictures on the screen, the rear wall of the projection booth should also be kept dark, so that reflected light from the screen which has passed through the observation port may not be returned to the screen.

To the management of a major cinema or TV studio, the review room is often the most important enclosure on the premises. The reason is that it is used by the producer, director, the cast, the camera, sound, and lighting crew plus publicity directors. This means that it should be able to provide the best visual and acoustic reproduce conditions possible. Thus, the screen luminance should well meet the American National Standards Association and the SMPTE standards of 16 ft-lamberts at the center of the screen. Since it is generally a motion picture theater equipment supply house which provides the screen and projection equipment, it should be made aware of the requirements in these rooms which, generally, are designed to accommodate between 25 and 40 people. Obviously, the interior decor of the enclosure should also be of a highly pleasing appearance, along with the upholstered chairs, the carpets, and other appointments.

Chapter 22

Future Sound-Recording Studios

It is not unusual nowadays to find beautiful recording studios in the country. They are generally near mountains or else along a lakeside. Such a location allows "superstars" to take out time for skiing, fishing, hunting, sailing, etc., particularly when the studio also has sleeping and living quarters.

When such a studio is in the mountains and its architecture is to conform to that of other buildings near or far, it may be advisable to construct the facilities in the form of an A-frame structure. However, a triangular room is undesirable acoustically because it results in a nonuniform distribution of the sound energy therein. Hence, the studio itself should be fitted with a suspended acoustic ceiling to achieve adequate low-frequency absorption in the enclosure.

But even though such an enterprise may be planned in the mountains, it is desirable to conduct a noise exposure level survey there, preferably over some months. An example of the desirability of such a noise level survey is illustrated in Fig. 22.1. The prospective studio site was located in the Sierra Nevada Mountains east of Fresno, California. At certain times of the year the area is plagued by tule fog, which forces aircraft to fly over a mountain route when, without the fog the planes could fly at a much lower altitude through a wide canyon. These are not scheduled commercial aircraft which fly at a fixed high elevation but air taxis between Mammoth Mountain ski resorts and the west coast, with a stopover in Fresno.

The owners of mountain and seaside studios often insist on the installation of either a number of small or a large picture window in the wall facing the most attractive view. Because the windows are not generally used during a recording session, they are covered with a curtain or a set of draperies, for the dual purpose of not distracting the performers during a recording and to prevent undesirable strong reflections or echoes from the hard panes of glass. This type of window covering should be folded or curled when it has been drawn over the window, so as not to generate pro-

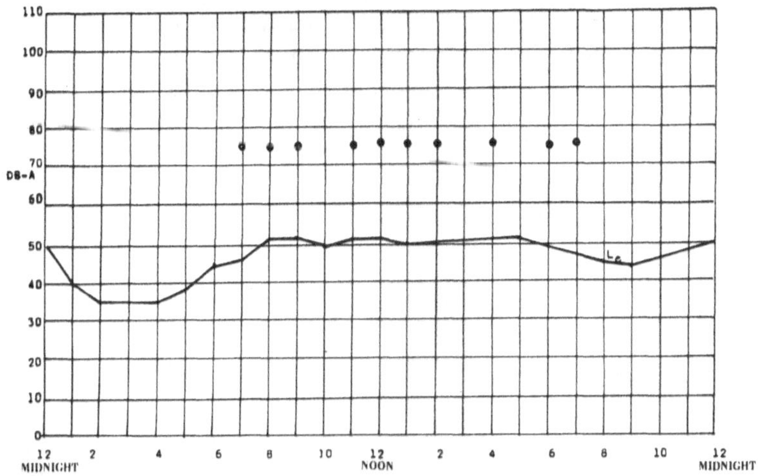

Fig. 22.1 Variation of A-weighted noise equivalent level over 24-h period in a mountainous region east of Fresno, California. Circles represent flyover disturbances of air-taxis forced to fly high on account of tule fog.

nounced dips and peaks in the absorption characteristic of the studio. This effect is discussed in greater detail later, when the effects of a cyclorama screen are under consideration.

While in the city the effects of wind and rain-struck trees are usually negligible, out in the open the noises so produced are not inconsiderable. For instance, when rain and wind strike a deciduous tree in full leaf, a noise level of 60 dB-A at 50 ft (15.2 m) from the tree is not uncommon for a wind velocity of 15 miles (24 km)/h and a rain drop velocity of 30 ft (9.1 m)/ s. Hence, when the internal noise level of the studio is to be kept no greater than 25 dB-A, the windows should exhibit a sound transmission loss of not less than 35 dB at 500 Hz. This undoubtedly calls for double-pane windows structurally isolated from each other as much as possible by surrounding the panes with a soft Neoprene extrusion, and by installing a generous amount of sound-absorbent material on the walls between the panes. The total surface density of the glass, for a minimum 5-in. (12.5 cm) pane separation, should not be less than 6 lb/ft^2 (29.3 kg/m^2).

Some urban studios, to induce clientele to come to their facilities, have provided a simulated outdoor scenery inside, as was done at Village Recorders.* Fig. 22.2 shows a Kodachrome slide projector set-up on the ceiling of the studio which can project a mountainside or seascape onto a large screen installed in an adjoining enclosure. The windows not only exclude any undesirable recording conditions about the screen, but also tend to provide a panoramic view, as if one were looking through the windows of a chalet.

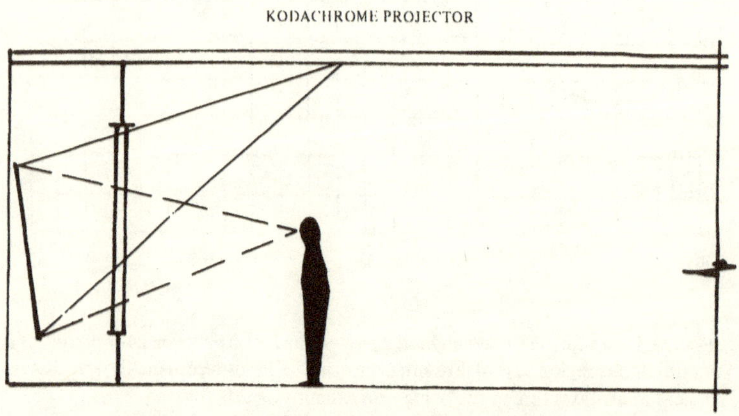

Fig. 22.2 Pictorial attempt to bring the outdoors indoors

At another studio which employed a high vaulted sound-absorbent ceiling it was possible to install many very small lights some of which could twinkle a little, to simulate a night sky when the lights in the room were turned off.

Many studios also provide recreational enclosures, complete with ping-pong tables, German draught beer, barbecue grills, etc. Others have saunas or a small pool, while still others feature large-screen television devices for reproducing video cassettes, or even live programs of soccer ball games or prize fights.

It's apparent that some studios will undoubtedly install TV tape machines for recording the band in picture and sound. This generally requires a cyclorama screen for direct photography or for back-screen projection of a landscape or mountains. For this reason a few words about cyclorama screens are in order.

---

* 1616 Butler Ave. in Los Angeles, CA.

A screen of this type, suspended at a distance from a wall, is a low-frequency absorber because it converts acoustic energy into mechanical energy by deforming and moving the flexible panel.

The resonance frequency of a cyclorama screen is given by the equation

$$f_0 = \frac{60}{(MD)^{1/2}} = \frac{170}{(md)^{1/2}}$$

where

M = surface density of screen (kg/m^2)
m = surface density of screen (lb/ft^2)
D = distance between screen and hard wall (m)
d = distance between screen and hard wall (in.)

A soft, more-or-less sound-transparent screen shows maximum absorption when the distance between the screen and the wall is a quarter wavelength and minimum absorption when the distance is a half wavelength. Similarly, odd multiples of the frequency of maximum absorption are also absorption peaks and even multiples of the frequency of minimum absorption are also absorption dips. For this reason, to smooth out the absorption characteristic, the cyclorama screen should not be placed parallel with the wall but at an angle, and the wall should carry sound-absorbent treatment, as shown in Fig. 22.3.

When it is not desirable to place the cyclorama screen at an angle to the walls, and when a heavily abosrbent wall is also not wanted, the walls should be corrugated as shown in Fig. 22.4. If it is further desired to achieve some acoustic variability for the boundaries behind the cyclorama screen, any of the devices illustrated in Fig. 22.5 may be employed.

Many existing studios may not have sufficient ceiling height to light a cyclorama screen adequately from the top without showing the luminaires. In such cases a deep groove or channel may be opened in the floor and the screen may be sunk into it to allow lighting it from below without showing the lights. At the aforementioned Osmond Studios the stage was so high (55 ft or 16.7 m) that the high cyclorama had to be lit both from the ceiling and from the floor pit. This arrangement also allowed air circulation from the middle of the studio to the wall registers because the screen was placed only 2 ft (0.6 m) into the 3-ft (0.92 m) deep channel. By this arrangement the air could pass under the screen to the walls.

Before going on with a discussion of future studio shapes, it may be desirable to note some apparently advantageous plans which, when tried,

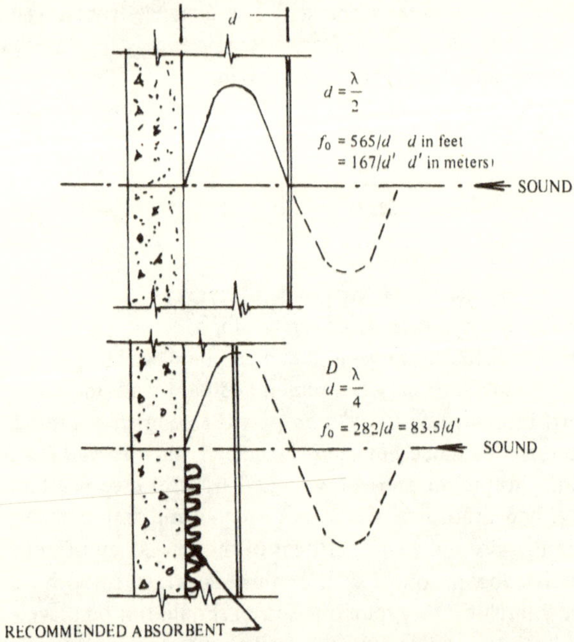

Fig. 22.3 Top diagram shows minimum absorption for cyclorama screen when it is a half wavelength from the hard wall, while the bottom diagram shows the maximum absorption for quarter wavelength distance. Sound-absorbent on wall reduces peaks and dips.

did not prove desirable. One such design is based on an experiment by Paul E. Sabine[1], who employed a number of pillars stationed about the walls of his reverberation chamber to achieve an increased diffusion of sound therein and a more nearly exponential sound decay after the signal was suddenly terminated. When such pillars were stationed about the walls of a large recording studio, the effect on the quality of the direct and the recorded music there was disastrous. In the words of one well-known musicologist, the renditions sounded as if they were produced in an "acoustic fog," that is, they were not clean, but lacked definition and "presence." This may be due to a slow sound decay, instead of the more normal decay characterized by an initial sharp drop in the signal, followed by a more prolonged decay. Fig. 22.6 shows a schematic plan of such a stage area.

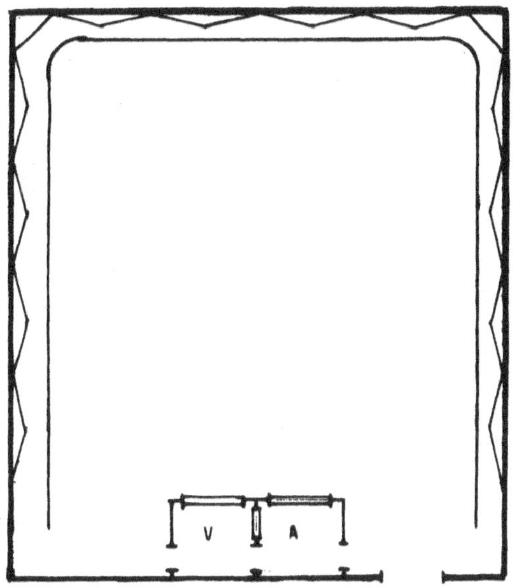

Fig. 22.4 Video disk recording studio with cyclorama screen and one audio and one video control room.

Another attempt to achieve increased sound diffusion in a studio consisted of the installation of helium-filled balloons below the ceiling. With helium being lighter than air, the balloons could simply float on the ceiling and represented no fire hazard unlike, say, hydrogen-filled balloons. With the speed of sound in helium 2.9 times that in air, the angles of refraction (for small angles of incidence) are 2.9 times the angles of incidence. Unfortunately this condition results in a critical angle of only 21 degrees. This means that for angles of incidence larger than 21 degrees, the sound is no longer refracted in the balloon but must either travel along the normal to the spherical surface or be reflected. Hence, only a small part of the incident signal can be so dispersed.

It may be interesting to note that, because of the large index of refraction of helium-filled containers, such bags have been used as portable noise barriers at construction sites. A small portion of the randomly incident din is transmitted through the gas-filled blankets, but most of it is reflected as if the barrier were made of heavy material.[2]

Fig. 22.7 shows a graphic illustration of sound striking a helium-filled balloon.

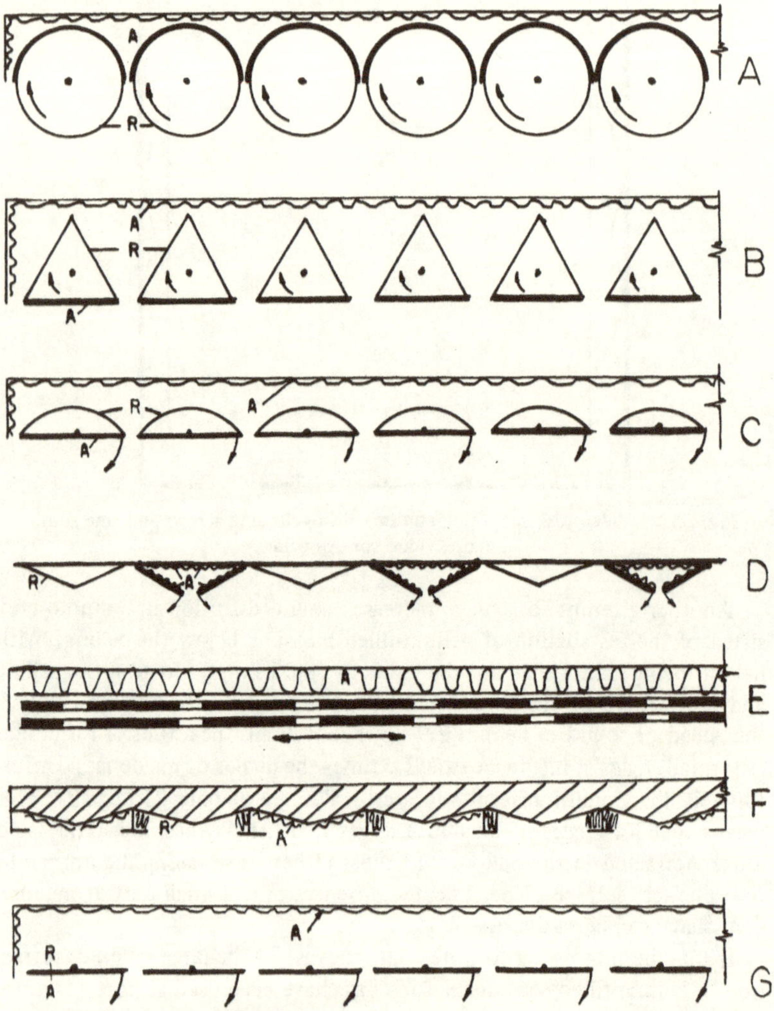

Fig. 22.5 Variable sound absorbers. A: rotatable cylinders; B: rotatable triangular prisms; C: rotatable cylindrical sections; D: folding splays; E: one stationary and one sliding perforated hardboard; F: retractable curtains; G: rotatable flat panels.

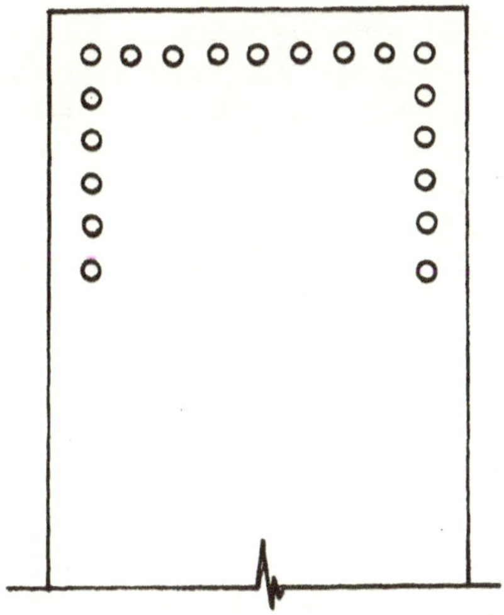

Fig. 22.6 Cylindrical pillars stationed about the stage area of a recording studio to increase the sound diffusion in it proved undesirable in that they caused a lack of definition of the instruments, generating the epitaph "acoustic fog."

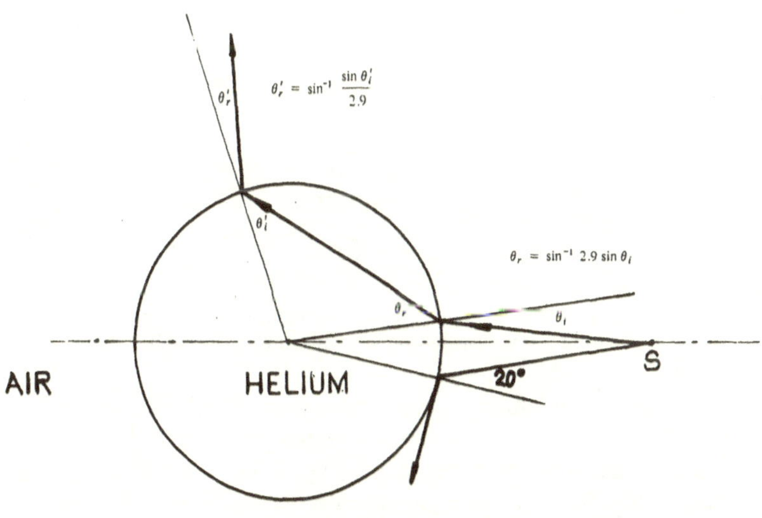

Fig. 22.7 Helium-filled balloons can refract sound only for incidence angles smaller than 20°.

A third novelty was a rotating helical column in a control room to smooth out the interference pattern in the enclosure. The device was approximately 2 ft in diameter and 6 ft tall, and turned 60 times per minute. So-called sound transmission tests made by reproducing sinusoidal tones from the monitor loudspeaker and recording them near the mixing console showed practically no effect in the transmission pattern when the unit was in a corner of the room, turning. Fig. 22.8 shows a schematic view of such a device.

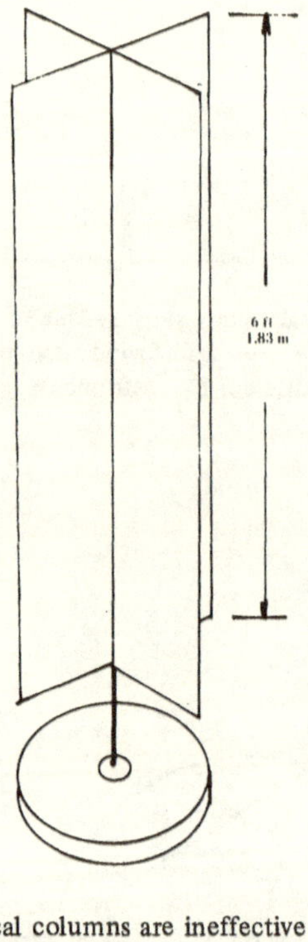

Fig. 22.8 Rotating vertical columns are ineffective in providing a more uniform sound distribution in a control room or studio.

Progress in technology does not mean that architectural acoustics can make similar progress and employ new shapes for a recording studio which before were not considered acceptable. Fig. 22.9 shows the sound-focusing pattern produced by a source in a circular room. A moderate amount of acoustic treatment on the walls of such a room has very little effect on the sound-concentrative "caustic" curve. This is because when the treatment

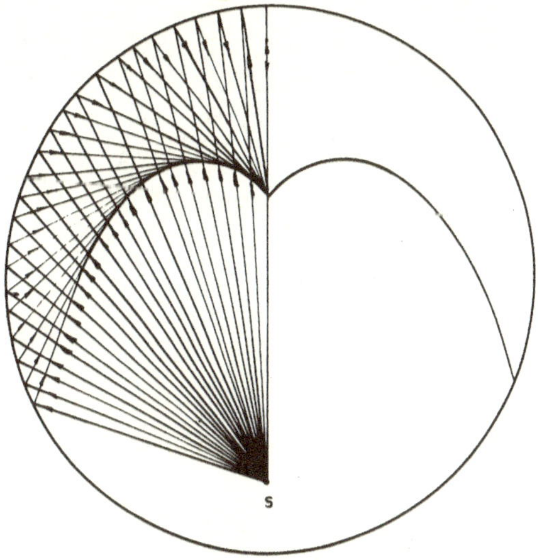

Fig. 22.9 Sound source in a circular room produces an arc of foci

is, e.g., 50 percent absorptive, the reflected sound energy density from such a wall is only 3 decibels below the direct energy density impinging on the surface. Even when the treatment is 90 percent absorptive, the reflected energy density is still only 10 decibels below the incident one. Because so many sound "rays" come to a focus along a line, the sound intensity along this line is objectionably higher than anywhere else in the enclosure. Similar focusing effects occur in polygonal rooms, shown in Fig. 22.10.

With TV programs in the offing in recording studios previously devoted to sound programs only, it become necessary to provide a video monitoring room in the studio. This may be "blind," that is, the program in the studio may be watched over closed circuit TV channels, or a sound-retardant window may be installed in the partition between the control room and the studio, so that the video mixer may watch the performers directly.

Some TV studios, as in the case of he Osmond Studios in Orem, Utah, have three control rooms, one video room in which the program director, sponsor, etc., may be stationed; a second TV control room for color adjustment, and a third enclosure devoted to sound mixing only.

With the audio control rooms probably shrinking in size to allow the installation of video monitoring booths, it becomes important to consider

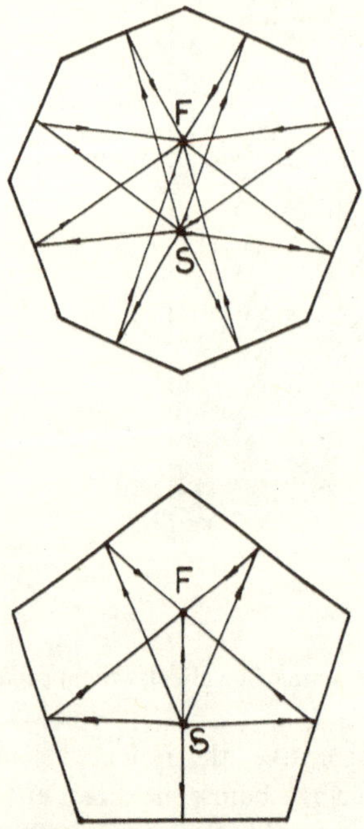

Fig. 22.10 Polygonal plans for a room do not prevent foci. S is sound source; F is sound focus

the ratio of the reflected sound energy density to the direct sound energy density to avoid interference effects.

When an observer is in the "far" sound field, where the reflected sound energy exceeds the direct energy, interference effects take place at prolonged notes, particularly at the low notes when the average sound absorptivity $\bar{a}$ in the room is small. In the "near"field the listener hears chiefly the signal reproduced in the manner of the loudspeaker response, because the reflected sound energy there is generally small.

As an example consider a 10 ft × 17 ft × 29 ft control room (not at all unusual), with a volume of 4930 ft^3, a total interior surface S of 1906 ft^2, a reverberation time of 0.5 s at 100 Hz, and a corresponding average sound absorptivity $\bar{a}$ of 0.23. When the mixer sits 10 ft from the loudspeakers, the ratio of reflected to direct sound energy density is given by

$$\frac{E_r}{E_d} = \frac{50D^2(1-\bar{a})}{aS} = \frac{50D'^2(1-\bar{a})}{aS'} \qquad \begin{array}{l} D' \text{ in meters} \\ S' \text{ in square meters} \end{array}$$

$$= \frac{50 \times 10^2(1-0.23)}{0.23 \times 1906}$$

$$= 8.8$$

The smaller the room is, reverberation time remaining constant, the shorter is the distance between sound source and observer where the reflected sound energy density equals the direct energy density. For this reason anechoic wedges appear desirable as the acoustic treatment of the rear wall of a control room. There are now 2-ft deep units on the market with a wire-net over them which does not affect their absorptivity but still protects them against prying fingers and curious hands.

Fig. 22.11 shows the variation of the ratio of reflected to direct sound energy density with room volume for various distances when the reverberation time in the rooms ins 0.5 s.

Fig. 22.12 shows plan of a studio intended for TV disk recording. Note the two control rooms, one of which may double for a video monitoring enclosure. Not shown are large-screen TV projection facilities which undoubtedly will have to be installed for reproducing a recording.

Fig. 22.13 shows the isolation measures employed in one studio to provide a floor with a constant elevation over the entire first floor. This is for the purpose of allowing handicapped persons in wheelchairs to move easily from one room to another without having to be wheeled up or down ramps. Since the office of this building complex contains a wide variety of electronic equipment, it was equipped with a computer floor the same as the adjoining control room. The stage, on the other hand, had a solid concrete floor treated with 3/16 in. thick epoxy which was water-tight and could be painted repeatedly with a watercolor paint without leaving a stain on the floor.

The latest trend in studio construction is the use of portable digitalized cinemobiles. A number of mobile analog recording studios have been constructed to allow star performers to enjoy the wilderness away from the noise and smoke of the cities, and digital video- and cinemobiles are the answer to the problem. A trailer of this type is preferably constructed with an external aluminum skin and internal plywood facing, and adequate acoustic treatment for both the control room and the studio proper. One

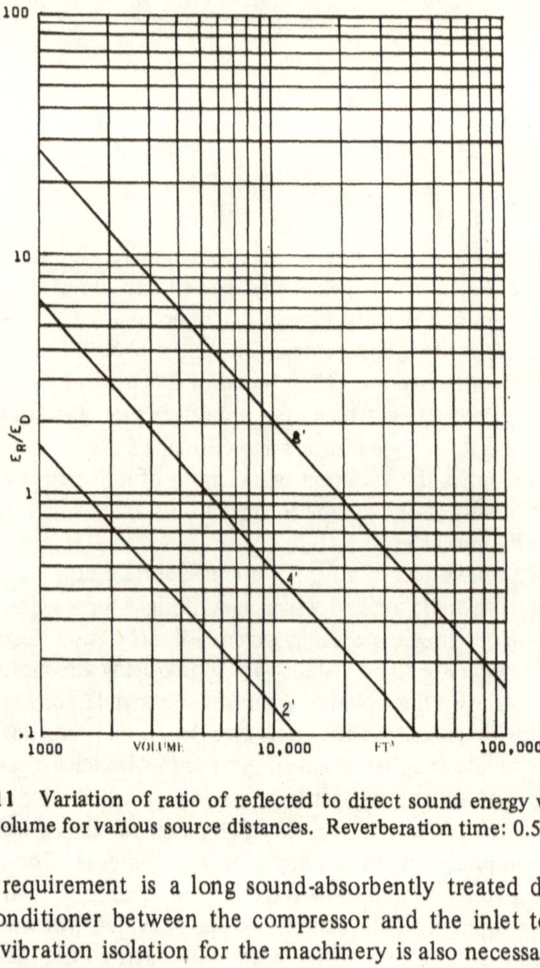

Fig. 22.11 Variation of ratio of reflected to direct sound energy with room volume for various source distances. Reverberation time: 0.5 s.

necessary requirement is a long sound-absorbently treated duct for the unit air conditioner between the compressor and the inlet to the room. Adequate vibration isolation for the machinery is also necessary. See Fig. 22.14.

There are fashions in music as much as there are in clothes. If the present trend is to record renditions in studios so dead that they resemble anechoic chambers, it may well be that in the future wood panels placed around the performer will be used to satisfy him as well as the customer of the disk or tape cassette. The tonal quality which has been missing in some multi-track recording rooms is what instrumentalists call attack, or attack time, the immediate tonal response produced by nearby reflectors.

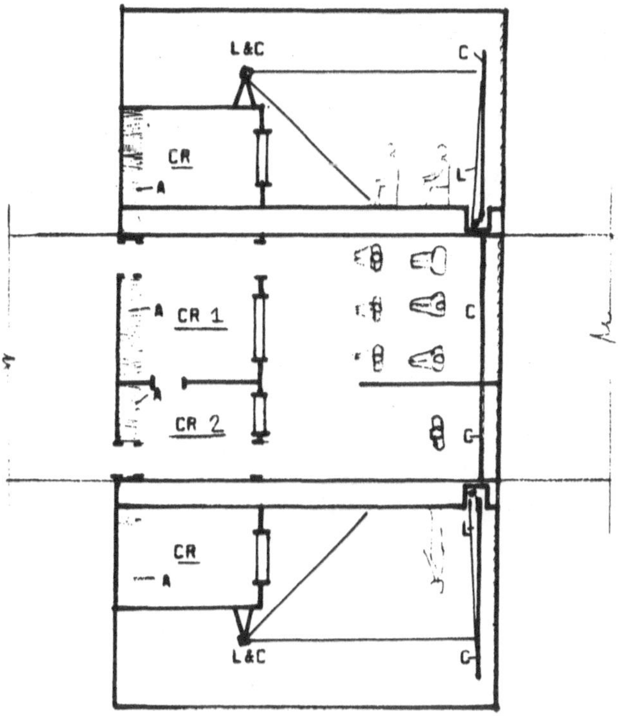

Fig. 22.12 Video recording studio. L & C: lights and camera; C: cyclorama screen; A: anechoic treatment; L: lights

Anticlimatically it may be said that a studio should please both the performers and the ultimate purchaser of the recorded piece of music.

The use of an attack time or early tonal reinforcement also tends to increase the definition of a note. G.W. Stewart,[3] in the early 1930's, was able to apply the well-known Heisenberg principle of uncertainty in physical quantities (wherefore one may never detect the location of an electron within an atom) to the acoustic case. His results, in brief, are stated by the equation $\Delta f = 1/\Delta t$, where f represents frequency; t, the time of observation, and Δ stands for class interval. As an example, to recognize a 1000 Hz tone within an error of 1 percent or 10 Hz, requires an observation period of 0.1 s. When a reflector is able to increase the time by 0.1 s, the error decreases to 0.5 percent.

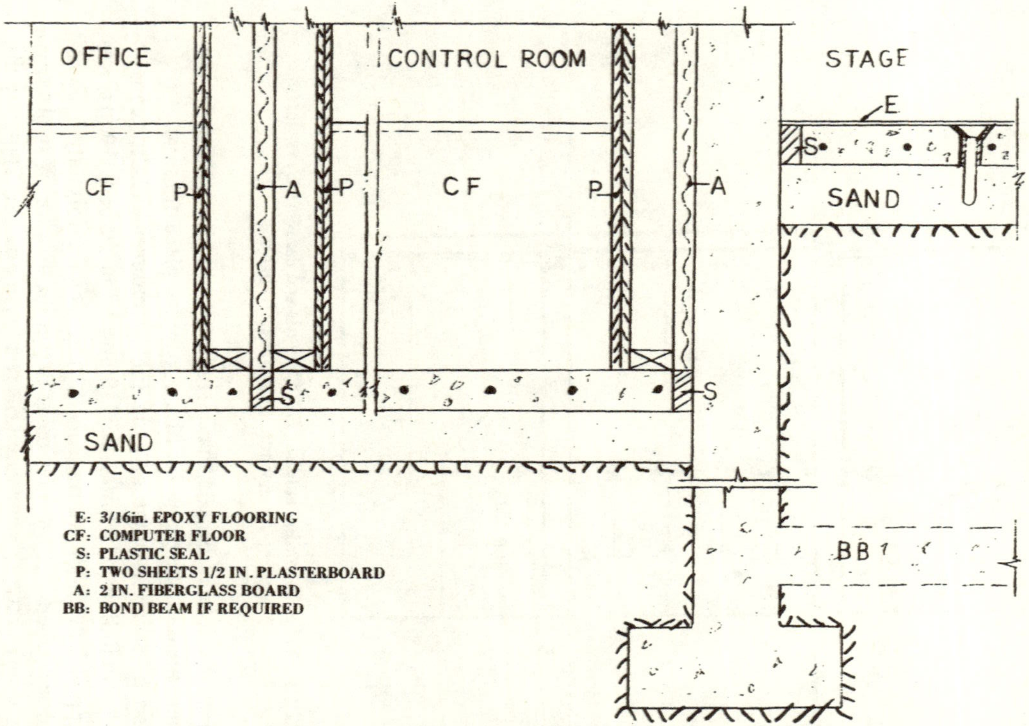

Fig. 22.13 Control room floor and studio floor have the same elevation to assist handicapped. E: 3/16 in. epoxy flooring; CF: computer floor; S: plastic seal; P: two sheets 1/2 in. plasterboard; A: 2 in. fiberglass board; BB: bond beam if required.

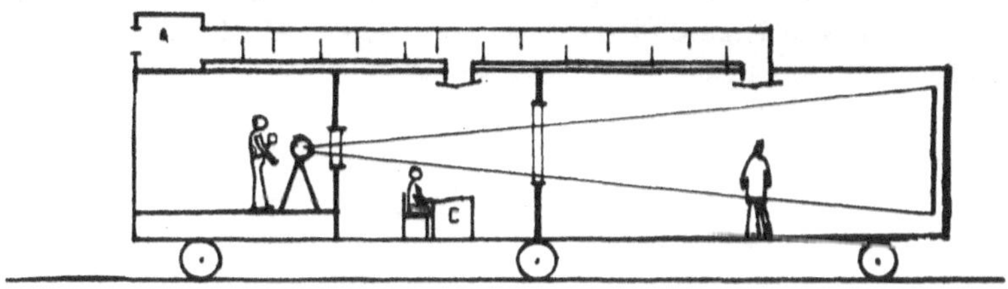

Fig. 22.14 Mobile TV recording studio with digitalized equipment. P: projector; A: unit air conditioner; C: mixing console

22.1 APPENDIX

The ratio of the reflected sound energy density to the direct energy density for a nondirectional source in a room is

$$\frac{E_R}{E_D} = \frac{16\pi D^2(1-\bar{a})}{aS} \cong \frac{50D^2(1-\bar{a})}{aS}$$

where

D = distance from source (ft or m)
$\bar{a}$ = average absorptivity
S = interior surface (ft^2 or m^2)

This expression can not be plotted in a highly illuminating manner, because $\bar{a}$ is generally not readily visualized, while the interior boundary surface, unlike the volume of a room and its reverberation period, is not well known. One may, however, make the substitutions $\bar{a} = 0.05V/ST$, where V is the volume of the enclosure and T is its reverberation time; and $V/S = 0.15V^{1/3}$. The latter expression is applicable to normally proportioned rooms, with ratios of dimensions in the order of 1:2:3, 2:3:5, etc. Then

$$\frac{E_R}{E_D} = \frac{1000D^2(T - 0.0075V^{1/3})}{V}$$

The above equation is shown graphically in Fig. 22.11, for the case when the control room reverberation time is 0.5 s.

When a vibrating panel is placed before a hard wall, resonance occurs when the mass reactance of the panel per unit area equals the compliance reactance per unit area of the trapped air. Thus

$$2\pi f M = \frac{\gamma P}{2\pi f D} = \frac{\rho c^2}{2\pi f D}$$

$$\text{Compliance}/m^2 = \frac{D}{\gamma P}$$

where

M = surface density of panel (kg/m²)
f = frequency
γ = ratio of specific heats
= 1.41
c = velocity of sound in air
= 344 m/s
D = depth of air space (m)
P = atmospheric pressure
= 10^5 N/m²
ρ = density of air
= 1.2 kg/m³

Solving the preceding equation for f,

$$f = \frac{1}{2\pi}\left(\frac{\gamma P}{MD}\right)^{1/2} = \frac{1}{6.28}\left(\frac{1.41 \times 10^5}{MD}\right)^{1/2} = \frac{60}{(MD)^{1/2}}$$

$$= \frac{c}{2\pi}\left(\frac{\rho}{MD}\right)^{1/2} = \frac{344}{6.28}\left(\frac{1.2}{MD}\right)^{1/2} = \frac{60}{(MD)^{1/2}}$$

$$= \frac{170}{(md)^{1/2}} \quad \text{if } m \text{ is in lb/ft}^2 \text{ and } d \text{ is in inches}$$

$$= \frac{60}{(m'd')^{1/2}} \quad \text{if } m' \text{ is in kg/m}^2 \text{ and } d' \text{ is in inches}$$

REFERENCES

1. Sabine, Paul E.: "Effects of Cylindrical Pillars in a Reverberation Chamber," J. Acoustical Society of America, V. 10; No. 1; July 1938. p 1.
2. Ailman, C.M. "A Light-Weight Helium-Filled Noise Barrier," INTERNOISE 78, p 513.
3. Stewart, G.W. "Problems suggested by an Uncertainty Principle in Acoustics," J. Acoustical Society of America, V. 2; No. 3; January 1931, p 325.

Chapter 23

Canopied Amphitheaters

The recording of music is not always done indoors, because often a famous touring band can not be made available in the studio. It has happened, therefore, that sound trucks have been sent to an outdoor entertainment facility for the recording of a "live" concert. However, some amphitheaters near a city may not be allowed to hold rock concerts because of the possible noise complaints from nearby neighbors. In such cases it may be possible to construct a canopy over the seating section of the theater to contain the sound better. A discussion of the design of a canopied amphitheater to permit rock concerts follows. Since nowadays TV is often involved in the recording of such outdoor concerts it appears desirable to construct the facility with adequate sound, video, and lighting control rooms. A canopied amphitheater is well suited to secure such requirements by building small enclosures below the canopy, out of the sight of the spectators. Such audio and video recordings could represent substantial income to a community or a private enterprise in control of the facilities. With the stage and the seating section adequately covered, the facility represents an all-weather show place, providing even more income to the owners.

Open-air amphitheaters are often the cause of peace disturbance complaints from residential areas adjoining the entertainment facility. The over-amplified music by the sound-reinforcement system tends to add to the annoyance charges. The situation can be made worse during the summer evenings when temperature inversion exists (temperature rising with elevation), because part of the sound power radiated skyward is able to return by refraction to distant locations about the theater.

The spectator section of the amphitheater under consideration consists of a sloping canopy or roof, a rear wall, and open or partially open sidewalls. The front part of the structure facing the stage is open also. Both the underside of the overhead cover and the rear wall carry a weatherproof and nondefaceable sound-absorbent treatment. This reduces radiations

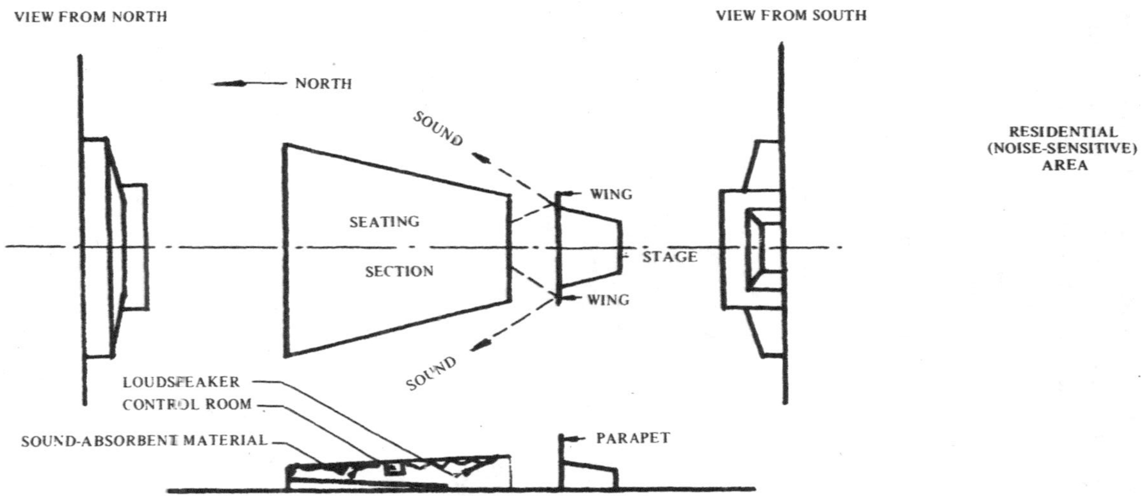

Fig. 23.1 Canopied grandstand or stadium for amphitheater to protect adjoining residences from sound level.

towards the stage by the time-delayed signal reproductions from the loudspeakers installed below the roof, as illustrated in Fig. 23.1.

Such a structure is able to reduce the outdoor transmission of the reproduced music by being contained to some extent within this "shed." This is made possible not only by employing a few low-positioned loudspeakers to cover the listeners adequately with signals, compared to blanketing them with reproduced sound from a stage-located emitter with a very high sound level at close range, but also because the sound radiated out of the enclosure suffers reflection by the acoustic mismatch between the interior of the seating area and the wide-open exterior about the facility, while considerable sound energy is absorbed by inter-reflections within simultaneously.

Fig. 23.2 shows plan and elevation of such a canopied amphitheater with a seating capacity of 1500 persons. Allowing 5 ft^2 (0.46 m^2) per spectator, exclusive of aisle space, the floor area comes to 7500 ft^2 (697 m^2), and with a 500 ft^2 (46 m^2) aisle space, the floor area comes to 8000 ft^2 (743 m^2), as shown in the figure.

Assuming 40 rows of seats 2.5 ft (0.76 m) distant from each other, each row rising 5 in. (12.5 cm) above the preceding one, the total floor rise is $40 \times 5/12 = 16.7$ ft (5.1 m).

The figure shows only six loudspeakers to cover a floor area of 8000 ft^2 (743 m^2). This is a very tentative estimate and was based on the following calculations. Assume that the radiation angle of the loudspeakers over most of the audio frequency range is 90 degrees. The circular coverage area on the floor is $3.14H^2$, where H is the slanting distance of the loudspeaker. If $H = 20$ ft (6.1 m), the number of loudspeakers is $8000/3.14 \times 20^2 = 6.7$.

For best sound coverage, "fill-in" units are sometimes used. A "fill-in" unit is like the center dot of a die showing five points.

For the loudspeaker arrangement shown on the figure, 3 sound-delay units are necessary. For instance, when the distance from seating area to stage is 113 ft, and sound requires 0.1 s to travel this distance, the delay should be set for 0.1 s.

It is difficult to calculate the reverberation time in the covered seating area, since all reverberation time formulas assume uniform sound distribution in a room, which is difficult to achieve in an enclosure with three open sides. From this investigator's experience, however, considerable sound absorptive treatment was necessary in the covered theater at Magic Mountain, California, to achieve satisfactory hearing conditions in the

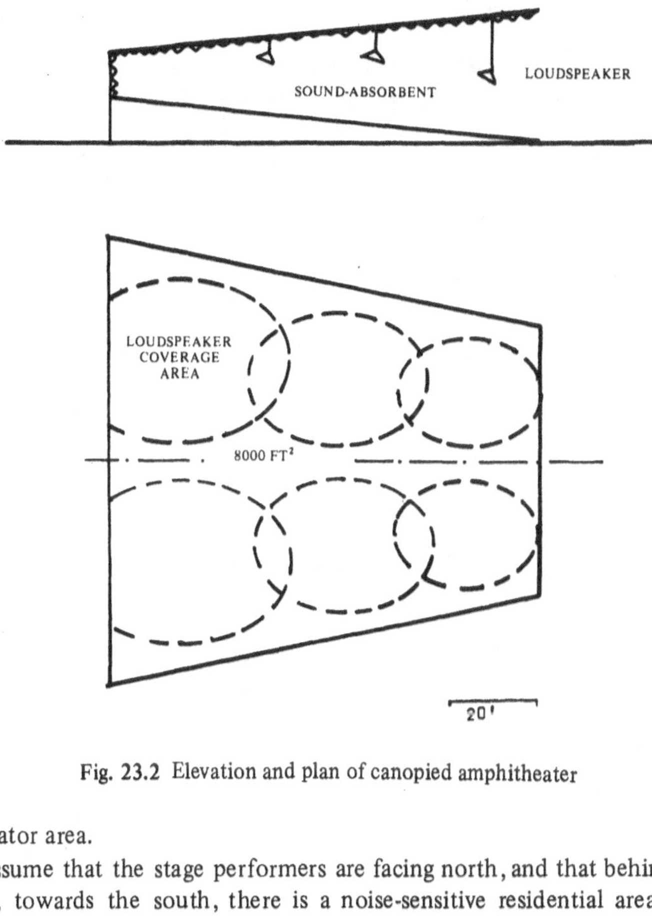

Fig. 23.2 Elevation and plan of canopied amphitheater

spectator area.

Assume that the stage performers are facing north, and that behind the stage, towards the south, there is a noise-sensitive residential area. By installing a parapet on the stage roof and mounting plywood, plastered "wings," or reflective panels on the sides of the stage, much of the sound radiating out of the seating section towards the stage will be reflected to the north by the parapet and the wings, thereby lowering sound transmission towards the south. These reflecting baffles should be treated sound-absorbently so as not to direct too much sound to the north, even though the area there is assumed to be less noise-sensitive (being zoned "commercial") than the south.

It is also desirable, if possible, to erect other structures, like buildings or even a concrete block wall, to the open sides of the canopied seating section, so as to minimize the transmission of sound in the easterly and westerly directions also.

The seats for the spectators should rise adequately towards the rear or north of the structure to allow a good vision of the activities taking place on the stage. To further reduce sound reflections in the seating section it is recommended that at least the aisles therein be covered with an indoor-outdoor carpet.

Both the rear wall and the roof should be substantially sound insulative to reduce sound transmission into the open. The roof may consist of metal decking. The rear wall may be either concrete blocks or stucco.

It should be noted that this type of construction allows the complete closing in of the canopied structure should the transmitted signal in the residential area be found objectionable later, either because of a noise level limit change or another reason.

Besides being able to reduce the sound transmission into adjoining residential areas, such a canopied amphitheater also has other advantages:

1. The seating area is shielded from overhead noises like aircraft fly-overs and helicopter overflights.
2. It allows the all-weather performance of concerts and other entertaining events.
3. By locating the control room below the roof, the mixer has a full view of the stage, can readily control the sound level within the seating area because he is in it, too, and the booth may also be used for lighting effects on the stage.

There is nothing new about canopied amphitheaters. One of the finest concert facilities in the world is the Tanglewood Music Shed in Lennox, Massachusetts. This incredibly beautiful outdoor music theater has an internal seating capacity of 6000, with an additional 6000 people being able to sit about the lawns outside the "shed." The acoustics of the structure are highly prized by many famous conductors and performers. Thus Isaac Stern, the famous violinist, considered it to have a brilliant, ringing sound with wonderful definition. In this well-known and highly publicized entertainment facility the lower portions of the sidewalls are open, while the upper ones carry 3/4 in. (1.9 cm) painted fiberboard. Since the "shed" is located in the Berkshire Hills, sound transmission to the far residence is no problem.

At Magic Mountain, in Valencia, near Los Angeles, there is also a canopied amphitheater with good acoustics. Being inside a private estate, sound transmission problems are also of little importance there.

From calculations made by this investigator as well as from measurements obtained from a scaled-down acoustic model of such a canopied amphitheater it has been learned that the mid-range or 500-Hz sound attenuation of such a unit is in the order of 25 dB, compared to an open-air, unbaffled facility with a high-pressure sound-reinforcement system in which one or two high-power loudspeakers are placed above the proscenium arch of the stage.

Fig. 23.3 is a pictorial representation of the A-weighted sound levels

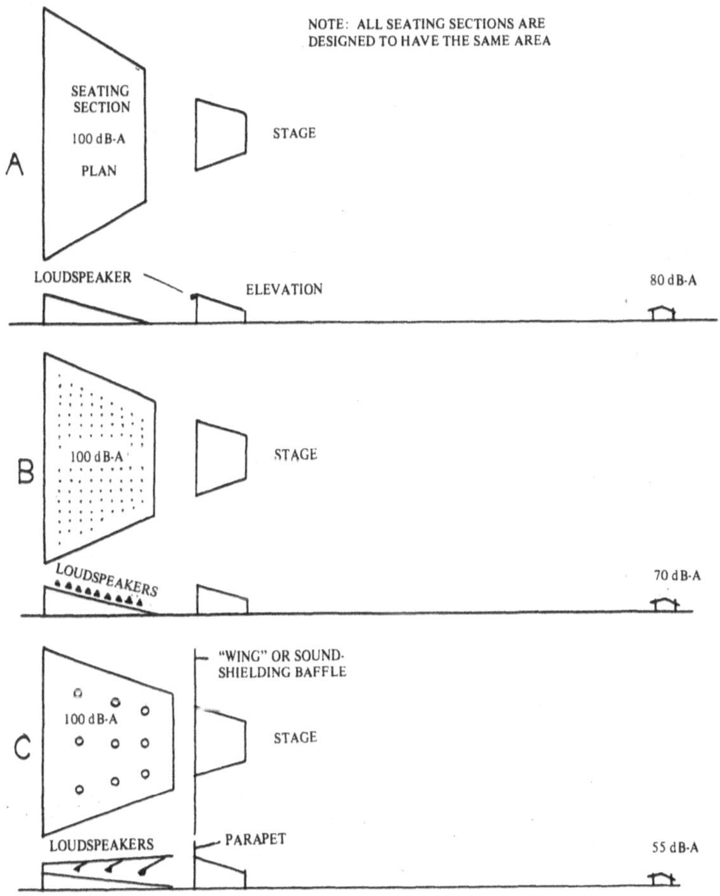

Fig. 23.3 Calculated A-weighted sound levels at 600 ft from the seating area of amphitheaters employing different sound-reinforcement systems in the open and in canopied unit, together with various types of sound baffles

which may be expected at 600 ft from the seating area of amphitheaters employing different sound-reinforcement systems in the open and in a canopied unit.

Diagram A on top of the figure shows the "conventional" amphitheater with a raised seating area and a high-pressure sound reinforcement system. The stage affords very little shielding because the rear elevation of the spectator section is practically the same as that of the stage, and the rear width of the seating area is too great for applying baffles to the sides of the stage. The noise level at the nearest neighbor, 600 ft (183 m) distant, is 80 dB-A.

Diagram B illustrates a low-pressure sound reinforcement system, employing many loudspeakers installed on 10-ft high posts. The estimated sound level at the nearest resident comes to 70 dB-A.

Diagram C represents a canopied amphitheater with wings to the sides of the stage and a parapet on top of the stage. The sound-reinforcement system within the enclosure is "hybrid" between the high- and the low-pressure systems. The noise level at the nearest resident is estimated to be 55 dB-A.

Fig. 23.4 shows recommended acoustic treatment for rear wall and roof of a canopied amphitheater.

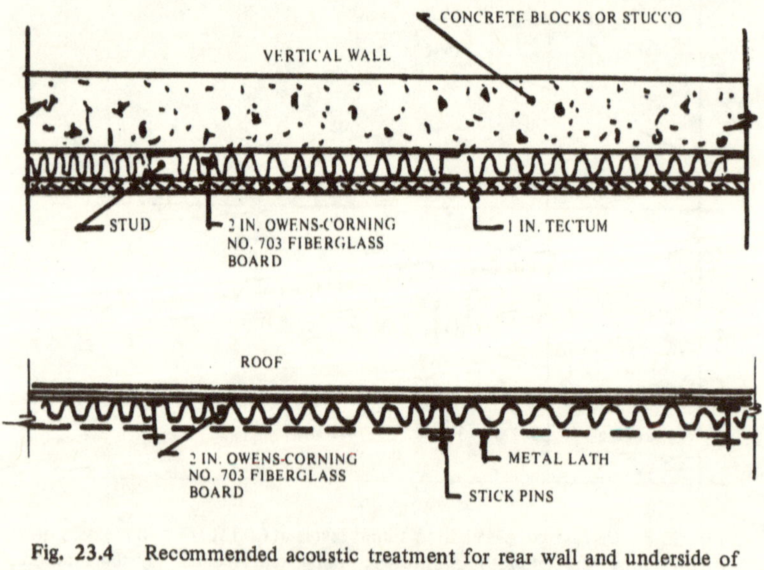

Fig. 23.4 Recommended acoustic treatment for rear wall and underside of roof of canopied amphitheater.

PART III

ELECTROACOUSTICS

Chapter 24

Microphone Properties

Before discussing microphone sensitivity and signal-to-noise ratios, it is important to establish some definitions. There are essentially four:

dBm – a measure of electric power referred to 0.001 W. A secondary reference is a given voltage squared divided by a given resistance, namely $0.7745^2/600 = 0.001$ W.

dBv – a measure of voltage referred to 0.7745 V rms.

dBV – a measure of voltage referred to 1 V rms.

S/N – Signal-to-noise ratio, which is the difference between the signal level and the noise level when *both* levels have the same reference quantity (volts, pressure, watts, etc.).

Considering that most microphone preamplifiers have an input impedance which is much higher than the output impedance of the associated microphone it appears reasonable to measure the microphone sensitivity in terms of its open-circuited output voltage for a given sound pressure at the transducer. To load a microphone with a resistance equal to its impedance reduces the signal-to-noise ratio at the preamplifier output, because of the lower input voltage. Such a resistance termination, or even the introduction of an attenuative network between microphone and preamplifier, is desirable only when substantial advantages are gained in other directions, such as avoiding overload in the amplifier, reducing distortion, flattening the frequency response of the system, etc.

The introduction of the MKS system of measurement has led to some confusion on the part of customers not familiar with microphone sensitivity ratings (there are several such systems). A lay person might have assumed that all microphones would be tested at the same sound pressure level, *SPL*, at the microphone, standardized by, e.g., the US Bureau of Standards. Unfortunately this is not so. Not only does the US Bureau of Standards not furnish standards in acoustics and electroacoustics, but there are several standardizing bodies in this country, like ANSI (American National Standards Institute), EIA (Electrical Industries Association), and the SMPTE

(Society of Motion Picture and Television Engineers), who shy away from a standard sound pressure level which could be used for a uniform microphone calibration.

In the older CGS system a microphone reference sound pressure (not reference sound pressure level) of 1 dyne/cm^2 was often used, which corresponds to an *SPL* of 74 dB referred to a minimum sound pressure at 1000 Hz (this minimum pressure is the same whether expressed in the old or the new system of metrication). In the MKS system, the preferred reference sound pressure was chosen as 1 N/m^2, which corresponds to an *SPL* of 94 dB. Hence the sensitivity of a microphone measured presently appears to be 20 dB higher than in the older days, but only because the *SPL* at the transducer is 20 dB higher (see Fig. 24.1).

In regard to the voltage generated by the microphone, it also requires a reference voltage if the rating is to be expressed in decibels. There are essentially two: 1 V and 0.7745 V, with corresponding decibels subscripts, respectively dBV and dBv.

In the EIA scheme of rating microphone sensitivity, which also employs the decibel unit, it is unimportant what the sound pressure is at the microphone. This is because in the term 20 log (E/P), the argument E/P remains constant: as the sound pressure P at the microphone increases, so does the microphone voltage E.

As an example of how to calculate the sensitivity of a microphone by the various proposed schemes, consider a small, low-priced unit, like a lapel microphone worn by TV announcers. At an *SPL* of 74 dB at the unit, the open-circuited voltage is 0.0001 V at its unterminated 150-ohm output terminal and 0.002 V at its 600-ohm unterminated terminal. For an *SPL* of 94 dB, the voltages are respectively 0.001 and 0.002 V.

1. For a 1 V reference, the microphone output levels are: (see Fig. 24.2)

		SPL	*Terminal* (Ω) 150	600
S	$= 20 \log E$			
	$= 20 \log 0.0001 =$	74	−80 dBV	
	$= 20 \log 0.0002 =$	74		−74 dBV
	$= 20 \log 0.001 =$	94	−60 dBV	
	$= 20 \log 0.002 =$	94		−54 dBV

Sound Pressure Level versus Sound Pressure

SPL dB	dynes/cm² μ bar	N/m² Pa	μPa
0	0.0002	0.00002	20
10	0.000631	0.0000631	63.1
14	0.001	0.0001	100
20	0.002	0.0002	200
30	0.00631	0.000631	631
34	0.01	0.001	1,000
40	0.02	0.002	2,000
50	0.0631	0.00631	6,310
54	0.1	0.01	10,000
60	0.2	0.02	20,000
70	0.631	0.0631	63,100
74	1.0	0.1	100,000
80	2.0	0.2	200,000
90	6.31	0.631	631,000
94	10.0	1.0	1,000,000
100	20.0	2.0	2,000,000
110	63.1	6.31	6,310,000
114	100.0	10.0	10,000,000
120	200.0	20.0	20,000,000
130	631.0	63.1	63,100,000
134	1,000.0	100.0	100,000,000
140	2,000.0	200.0	200,000,000
150	6,310.0	631.0	631,000,000
154	10,000.0	1,000.0	1,000,000,000
160	20,000.0	2,000.0	2,000,000,000
170	63,100.0	6,310.0	6,310,000,000
174	100,000.0	10,000.0	10,000,000,000
180	200,000.0	20,000.0	20,000,000,000
190	631,000.0	63,100.0	63,100,000,000
194	1,000,000.0	100,000.0	100,000,000,000

ρ = density of air
= 0.00118 g/cm^3
= 1.18 kg/m^3

c = sound velocity in air
= 34,400 cm/s
= 344 m/s

c^2 = 1.185 × 10^9 (cm/s)^2
= 1.185 × 10^5 (m/s)^2

ρc = 40.6 cgs rayls
= 406 mks rayls

ρc^2 = 1.4 × 10^6 g/(cm,s^2)
= 1.4 × 10^5 kg/(m,s^2)

1 Pa = 1 N/m^2
= 10 μbar
= 10 dynes/cm^2

1 μbar = 0.1 N/m^2
= 0.1 Pa
= 1 dynes/cm^2

Fig. 24.1

2. For the electrical power level relative to 0.001 W delivered by the microphone into a resistance equal to its own impedance:

	SPL	Terminal (Ω) 150	600
$S' = 10 \log \dfrac{(E/2)^2/R}{0.001}$			
$= 10 \log \dfrac{E^2}{R} + 24$			
$= 10 \log \dfrac{0.0001^2}{150} + 24 =$	74	−77.76 dBm	
$= 10 \log \dfrac{0.0002^2}{600} + 24 =$	74		−77.76 dBm
$= 10 \log \dfrac{0.001^2}{150} + 24 =$	94	−57.76 dBm	
$= 10 \log \dfrac{0.002^2}{600} + 24 =$	94		−57.76 dBm.

3. As noted previously, by the EIA system of rating microphone sensitivity it is unimportant what the sound pressure is at the microphone, because a higher sound pressure results in a higher open-circuited microphone voltage, so that the ratio of E/P remains constant; the rating varies, however, with the microphone output impedance.

For our sample microphone

$$G_m = 20 \log (E/P) - 10 \log R - 50$$

$$= 20 \log (0.0001)/1 - 10 \log 150 - 50 \quad = -154.76 \text{ dB at } 150\ \Omega$$

$$= 20 \log (0.0001)/1 - 10 \log 600 - 50 \quad = -151.76 \text{ dB at } 600\ \Omega$$

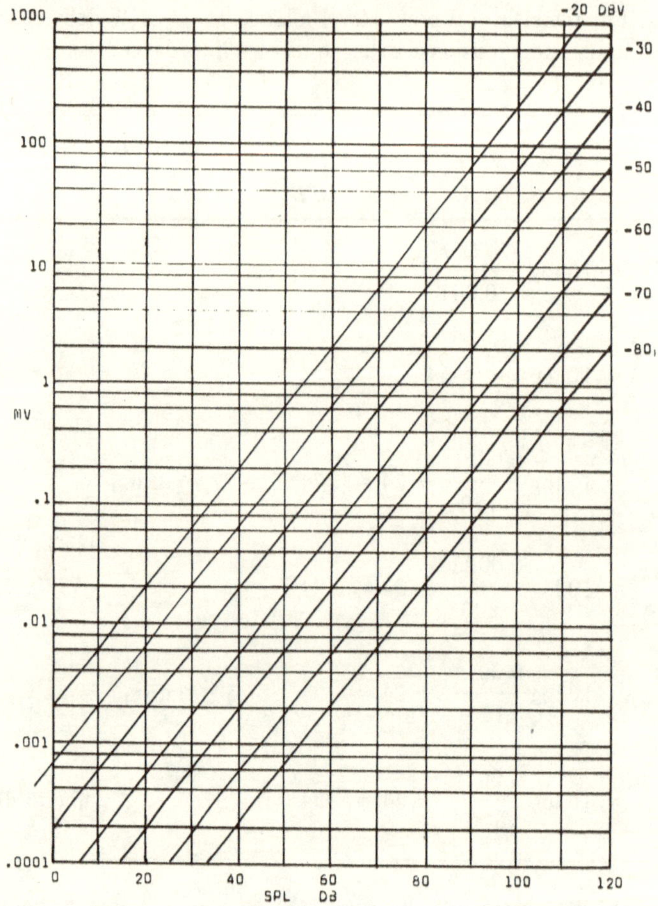

Fig. 24.2 Variation of microphone open-circuit voltage in millivolts, mV, with sound-pressure level, *SPL*, for various microphone sensitivity levels in dBV, referred to 1 V output for an MKS reference pressure of 1N/m²

When the open-circuited output voltage of the microphone at its 150-ohm terminal is –80 dB in reference to 1 V for a *SPL* of 74 dB, then when *SPL* is zero, the open-circuited output voltage becomes $-80 + (-74) = -154$ dB in reference to 1 V. This is equivalent to a voltage x of $20 \log x = -154$, or $x = 2 \times 10^{-8}$ V. At the 600-ohm terminal $x = 4 \times 10^{-8}$ V when *SPL* is zero, corresponding to an output level of –151 dB. Note that these level values practically equal the EIA ratings, since they also pertain to the sound pressure threshold at 1000 Hz, according to EIA Standard SE-105.

Closely associated with the sensitivity of a microphone is the signal-to-noise ratio which exists at the input to the associated preamplifier as well as at its output. The overall microphone noise voltage should preferably lie below the lowest noise voltage generated by an equivalent resistor.

The thermal noise voltage produced by the electrical resistance of a sound source is dependent on the frequency bandwidth under consideration, the magnitude of the resistance, and the temperature existing at the time of the measurement. This voltage is

$$E_n = (4kTRW)^{1/2}$$

where

k = Boltzmann's constant
 = 1.38×10^{-23} J/K
T = absolute temperature
 = 273 + room temperature (both in degrees Celsius)
R = resistance, ohms
W = bandwidth (Hz)

When $R = 600\ \Omega$, $W = 20{,}000$ Hz, $T = 300$ K, the white noise voltage across the source resistance turns out to be $E_n = 4.48 \times 10^{-7}$ V according to the above equation.

For a reference voltage of 0.7745 V rms across 600 Ω, the noise level voltage level equals

$$E_n' = 20 \log \frac{4.48 \times 10^{-7}}{0.7745}$$

$$= -124.8 \text{ dBv}$$

When the microphone output voltage is 0.0002 V across its unterminated 600-Ω terminal for an *SPL* of 74 dB, the corresponding voltage level referred to 0.7745 V is

$$S' = 20 \log \frac{0.0002}{0.7745}$$

$$= -71.76 \text{ dBv}$$

The signal-to-noise ratio for this case becomes

$$S/N = -124.8 + 71.76$$
$$= 53 \text{ dB}$$

This means that the lowest *SPL* at 1000 Hz which the unit can measure is 74 − 53 = 21 dB. When no overload occurs in either the microphone or the amplifier, the signal-to-noise ratio for the microphone is greater for higher signal levels. For an *SPL* of 121 dB at the microphone, the signal-to-noise ratio becomes 100 dB.

In 1965 this writer developed the RCA MI-10006A microphone which has an open-circuited output voltage of 0.004 V at its 600-Ω terminal for an *SPL* of 74 dB, which makes it 26 dB more sensitive than the small commercial unit discussed above. Its corresponding signal-to-noise ratio is 79 dB, and it is capable of measuring acoustic noise levels 5 dB below the 1000-Hz threshold value of zero dB.

It should be noted, however, that the electrical resistance noise of a microphone may not be the highest disturbance which limits the signal-to-noise ratio for a microphone of given sensitivity. This is particularly true for condenser microphones which may emit a "frying" noise when water vapor is able to collect between the diaphragm and the back electrode.

Chapter 25

Loudspeaker Sensitivity

Even a caual perusal of loudspeaker catalogs quickly demonstrates that practically no manufacturer employs the standards set by standardizing bodies like EIA, the Electronic Industries Association, or its predecessor the RTMA, the Radio Television Manufacturers Association, and others. The general excuse is that these standards are simply too complicated and involved. A description of the "matching" transformer between amplifier and loudspeaker alone (in RS-159) requires four pages of details, including its frequency response, transformer loss, power handling capacity, allowable distortion, and other specifications, plus the note that "in many sound fields the sound pressure ratios are not proportional to the square root of the corresponding power ratios and hence cannot be expressed in decibels in the strict sense."

Loudspeaker manufacturers, therefore, have chosen their own standards for expressing the sensitivity of their products. One such manufacturer employs a 10-ft distance from the horn of his unit to measure the pressure sensitivity while feeding pink noise into the transducer. The measurement is made along the longitudinal axis of the device while 1 W of electric power is directed to the reproducer. Another manufacturer, rating the sensitivity of his three-way system, prefers to publish the sensitivity of each of his three loudspeakers separately. For the measurement of the woofer it is stated that "since the major portion of the energy reproduced by the low-frequency loudspeaker lies below 800 Hz, this specification has been developed using a test signal warbled from 100 to 500 Hz rather than the conventional 1-kHz test signal on which the EIA sensitivity is based. With respect to the sensitivity of the mid-range speaker it is noted that the test result is averaged from 1 to 3 kHz within 1 dB. The tweeter sensitivity, similarly, is averaged above 2 kHz within 1 dB.

It is clear, therefore, that to learn of the pressure sensitivity of a loudspeaker one must read the catalog of the pertinent manufacturer.

Even less information is available about the sound power rating of a loudspeaker, which obviously entails an even more complicated test set-up than for the sound pressure sensitivity.

Chapter 26

Electronic Light Dimmers

It has become practice in the theater and the television industry to employ so-called *SCR* (Silicon Controlled Rectifier) lighting dimmers. They present numerous advantages to the user, over manually controlled variable autotransformers and resistance dimmers. A small amount of signal power at the input of the remotely located electronic dimmer station produces, through distant power amplifiers, full power to the lamps on the stage. Thus milliwatts of signal power can control kilowatts of lamp load instantly, unlike motor-positioned variacs with their inherent time lags. Also, the whole lighting set-up can be programmed by means of memory systems containing large quantities of cues. Indeed hundreds of dimmers can be so controlled in an elaborate functional program.

However, there are several disadvantages connected with electronic dimmers. Foremost is the lopsided waveform of the current which causes the radiation of objectionable harmonics. One should be aware here that the word rectifier in the term *SCR* is somewhat of a misnomer, in that the thyristers in inverse parallel do not produce a direct current. Rather, power change to the lamps is effected by interrupting the signals to the gates of the thyristers, and thus causing an interrupted current to the lamps. Because the rate of current change is in the order of 100 times a second, the light sources are not extinguished–the thermal inertia of the filaments maintaining the illumination. There is not even a flicker in the light output, only a reduction in the light intensity when smaller current pulses are sent to the luminaires.

The high currents in the cables placed in a steel raceway can cause the metal panels to vibrate by magnetic induction. To reduce this effect, the panels should be treated with a viscoelastic compound, and the duct itself should be mounted resiliently to its support. The use of aluminum, copper, or nonmagnetic steel for the raceway material greatly reduces the vibration, but by the same token provides little magnetic shielding for near microphones and video cameras. When maximum electric and magnetic shielding is required, the raceways may effectively be placed in plasterboard tunnels, as shown in Fig. 26.1. Put elastic supports at regular

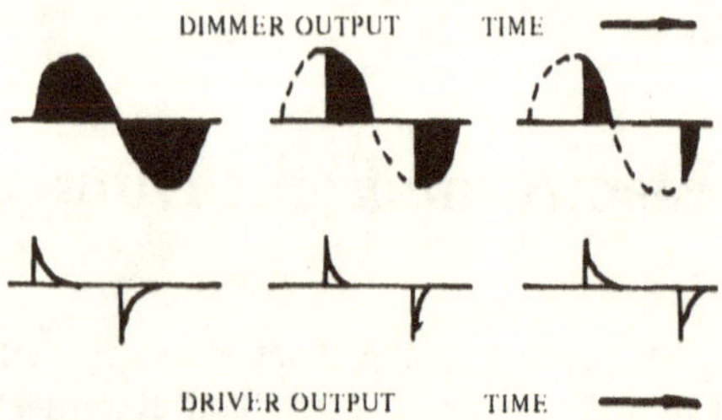

CURRENT WAVEFORM

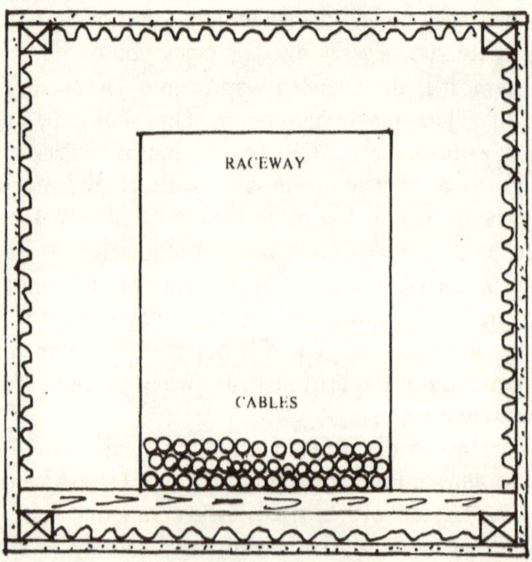

Fig. 26.1 Sound-absorbently lined plasterboard tunnel for steel raceway vibrating by magnetic induction from high-current electronic dimmer cables

intervals under the raceways, treat the inside of the tunnel with a 2-in. (5 cm) thick fireproof sound absorbent, and provide small openings in the tunnel every 10 ft so that ready access to the raceway can be had. This is the best and surest cure of such noise within a stage, because electric filtering of the pulsating current is only of limited effect.

For noises generated at the luminaires, two suggestions are made. Order the incandescent lights with extra supports for the filaments, so that they may not be able to vibrate so strongly, and treat "barndoors" and other lamp housing reflectors with a viscoelastic compound to prevent them from vibrating when they are made of a magnetic material, or else employ nonmagnetic reflectors.

The subject is one which properly belongs to lighting equipment manufacturers, and this investigator has asked the cooperation of the General Electric Co.*

The "electronic" cure of the problem consists in lengthening the rise time of the pulse wavefronts, usually from 5 μs to 500 μs. This is generally achieved by placing large inductors in series with the current, and sometimes by also placing condensers across the line to attenuate the high-frequency harmonics.

Fig. 26.1 also shows the current waveform with its irregular pattern, for which it is difficult to design an effective filter. Electronic dimmers often produce r-f interference which, of course, can to some extent be reduced by appropriate filtering at the input of the amplifiers used on the stage or in the video and audio control rooms. The use of acoustic filters near the raceways or near the microphones is completely impractical because of the wide frequency range comprised by the noise components.

One company well acquainted with the problems of electronic dimmers and means for their noise attenuation is the "Skirpan Lighting Control Corporation."**

It may happen, as in career planning, that axioms and dicta often fail to meet a specific case. There are also frequently conflicting requirements, for example, when an architect, for the given volume of a concert hall, prefers a low ceiling for the enclosure because such a room allows a greater floor seating capacity; while an acoustician may prefer a high ceiling for

* Mr. C.N. Clark, Marketing Planner of Stage/Studio and Audio-Visual Lamps, of the G.E. Lamp Marketing Department, Nela Park, Cleveland, Ohio, 44112, Tel: 216-266-2092.

** 61-03 32nd Ave., Woodside N.Y. 11377; 5444 Agnes Ave., North Hollywood, California (Tel: 877-0131).

the hall because of the resulting preponderance of lateral reflections favoring acoustic perspective and stereo sound reception. Neither designer is right or wrong—delivering the product in an atmosphere of compromise and open-mindedness is essential.

INDEX

www.ingramcontent.com/pod-product-compliance
Lightning Source LLC
LaVergne TN
LVHW090320210926
848480LV00001B/312